STATE DEVOLUTION IN AMERICA

URBAN AFFAIRS ANNUAL REVIEWS

A series of reference volumes discussing programs, policies, and current developments in all areas of concern to urban specialists.

SERIES EDITORS

David C. Perry, *State University of New York at Buffalo*
Sallie A. Marston, *University of Arizona*

The **Urban Affairs Annual Reviews** presents original theoretical, normative, and empirical work on urban issues and problems in regularly published volumes. The objective is to encourage critical thinking and effective practice by bringing together interdisciplinary perspectives on a common urban theme. Research that links theoretical, empirical, and policy approaches and that draws on comparative analyses offers the most promise for bridging disciplinary boundaries and contributing to these broad objectives. With the help of an international advisory board, the editors will invite and review proposals for **Urban Affairs Annual Reviews** volumes that incorporate these objectives. The aim is to ensure that the **Urban Affairs Annual Reviews** remains in the forefront of urban research and practice by providing thoughtful, timely analyses of cross-cutting issues for an audience of scholars, students, and practitioners working on urban concerns throughout the world.

INTERNATIONAL EDITORIAL ADVISORY BOARD

Swapna Banerjee-Ghua, *Lakmanya Tilak Bhawan University (Bombay)*
Robert Beauregard, *New School for Social Research*
Eugenia Birch, *Hunter College*
Sophie Body-Gendrot, *Université Sorbonne*
M. Christine Boyer, *Princeton University*
Susan E. Clarke, *University of Colorado at Boulder*
Adelia de Souza, *Universidade de São Paulo*
Sabina Dietrick, *University of Pittsburgh*
Arturo Escobar, *University of Massachusetts–Amherst*
Susan Fainstein, *Rutgers University*
Luis R. Fraga, *Stanford University*
Robin Hambleton, *University of the West of England*
David W. Harvey, *Johns Hopkins University*
Dennis R. Judd, *University of Missouri at St. Louis*
Paul Knox, *Virginia Polytechnic and State University*
Margit Mayer, *Freie Universität Berlin*
Michael Preston, *University of Southern California*
Neil Smith, *Rutgers University*
Daphne Spain, *University of Virginia*
Clarence Stone, *University of Maryland*

RECENT VOLUMES

34 **ECONOMIC RESTRUCTURING AND POLITICAL RESPONSE**
   Edited by Robert A. Beauregard

35 **CITIES IN A GLOBAL SOCIETY**
   Edited by Richard V. Knight and Gary Gappert

36 **GOVERNMENT AND HOUSING: Developments in Seven Countries**
   Edited by Willem van Vliet–– and Jan van Weesep

38 **BIG CITY POLITICS IN TRANSITION**
   Edited by H. V. Savitch and John Clayton Thomas

40 **THE PENTAGON AND THE CITIES**
   Edited by Andrew Kirby

41 **MOBILIZING THE COMMUNITY: Local Politics in the Era of the Global City**
   Edited by Robert Fisher and Joseph Kling

42 **GENDER IN URBAN RESEARCH**
   Edited by Judith A. Garber and Robyne S. Turner

43 **BUILDING THE PUBLIC CITY: The Politics, Governance, and Finance of Public Infrastructure**
   Edited by David C. Perry

44 **NORTH AMERICAN CITIES AND THE GLOBAL ECONOMY: Challenges and Opportunities**
   Edited by Peter Karl Kresl and Gary Gappert

45 **REGIONAL POLITICS: America in a Post-City Age**
   Edited by H. V. Savitch and Ronald K. Vogel

46 **AFFORDABLE HOUSING AND URBAN REDEVELOPMENT IN THE UNITED STATES**
   Edited by Willem van Vliet ––

47 **DILEMMAS OF URBAN ECONOMIC DEVELOPMENT: Issues in Theory and Practice**
   Edited by Richard D. Bingham and Robert Mier

48 **STATE DEVOLUTION IN AMERICA: Implications for a Diverse Society**
   Edited by Lynn A. Staeheli, Janet E. Kodras, and Colin Flint

STATE DEVOLUTION IN AMERICA

◆

IMPLICATIONS
FOR A
DIVERSE
SOCIETY

EDITED BY
LYNN A. STAEHELI
JANET E. KODRAS
COLIN FLINT

**URBAN
AFFAIRS
ANNUAL
REVIEWS
48**

SAGE Publications
International Educational and Professional Publisher
Thousand Oaks London New Delhi

Copyright © 1997 by Sage Publications, Inc.

All rights reserved. No part of this book may be reproduced or utilized in any form or by any means, electronic or mechanical, including photocopying, recording, or by any information storage and retrieval system, without permission in writing from the publisher.

For information:

SAGE Publications, Inc.
2455 Teller Road
Thousand Oaks, California 91320
E-mail: order@sagepub.com

SAGE Publications Ltd.
6 Bonhill Street
London EC2A 4PU
United Kingdom

SAGE Publications India Pvt. Ltd.
M-32 Market
Greater Kailash I
New Delhi 110 048 India

HT
108
.U7
vol. 48

Printed in the United States of America

Library of Congress Cataloging-in-Publication Data

ISSN: 0083-4688
ISBN: 0-7619-0879-X (Hardcover)
ISBN: 0-7619-0880-3 (Paperback)

This book is printed on acid-free paper.

97 98 99 00 01 02 03 10 9 8 7 6 5 4 3 2 1

Acquiring Editor:	Catherine Rossbach
Editorial Assistant:	Kathleen Derby
Production Editor:	Sherrise M. Purdum
Production Assistant:	Lynn Miyata
Typesetter:	Rebecca Evans
Print Buyer:	Anna Chin

Contents

Series Editors' Introduction ix

Acknowledgments xi

Introduction xii
Lynn A. Staeheli, Janet E. Kodras, and *Colin Flint*

Part I: Sources of State Change

1. State Restructuring, Political Opportunism,
   and Capital Mobility 3
   Robert W. Lake

2. Economic Globalization and Income Inequality
   in the United States 21
   John O'Loughlin

3. Globalization and Social Restructuring of the American
   Population: Geographies of Exclusion and Vulnerability 41
   Janet E. Kodras

4. Citizenship and the Search for Community 60
   Lynn A. Staeheli

Part II: Transforming the Political Opportunity Structure

5. Restructuring the State: Devolution, Privatization,
   and the Geographic Redistribution of Power and
   Capacity in Governance 79
   Janet E. Kodras

6. How Federal Cutbacks Affect the Charitable Sector 97
   Julian Wolpert

7. State Restructuring and the Importance of "Rights-Talk" 118
   Don Mitchell

Part III: Implications of State Change

8. FAIR or Foul? Remaking Agricultural Policy
   for the 21st Century 141
   Brian Page

9. Back to the Future in Labor Relations:
   From the New Deal to Newt's Deal 161
   Andrew Herod

10. Responsibility, Regulation, and Retrenchment:
    The End of Welfare? 181
    Meghan Cope

11. Transnationalism, Nationalism, and International
    Migration: The Changing Role and Relevance
    of the State 206
    Richard Wright

12. Education Policy and the 104th Congress 221
    Fred M. Shelley

13. Environmental Policy and Government Restructuring ... 233
    Marvin Waterstone

14. Conclusion: Regional Collective Memories and the
    Ideology of State Restructuring 252
    Colin Flint

    Index 272

    About the Contributors 283

Series Editors' Introduction

Equal amounts of concern and interest have emerged recently over the "devolution" of government in the United States. For some, this characterization of political and policy initiatives has meant change in the programmatic structure of government—as in its "reinvention." For others, devolution implies even more profound structural implications—including alternations in the ideological direction of the state and the alteration of socio-political relations of power. In either case, the implications of devolution—especially at the federal level—are unfolding all around us as some federal programs are devolved to the state or local levels, while others are privatized and still others threatened with outright elimination.

Such dynamic institutional impulses demand close attention. Therefore, we are pleased to have selected this new and timely collection of essays on state restructuring as one of the two Urban Affairs Annual Reviews volumes for 1997. *State Devolution in America: Implications for a Diverse Society* covers the range of federal responsibilities from agriculture to welfare, as well as addresses the increasing emphasis on the private sector. The book provides an excellent overview of how current restructuring is redefining the roles of all levels of government. Using the "Contract with America" as a starting point, the various contributors to *State Devolution in America* provide a critical analysis of how devolving federal responsibilities to city and state governments (and the shifting of federal functions to a highly fragmented private sector) create, accentuate, and in many cases exacerbate geographic variations in government provision. As the editors of the volume observe, a redistribution of the benefits and burdens of the federal government to lower levels of government translates into a redistribution of power and resources among and between cities and regions. One possible effect of such a redistribution is the escalating of interstate and interurban competition and conflict as cities and states are no longer able to turn to the federal "referee." The collection also shows how the political philosophy surrounding federal devolution is changing expectations and creating new relationships between people and government.

In focusing on the federal hierarchy, *State Devolution in America* departs substantially from the usual local and urban-centered contributions to this series. We are pleased to include the book in the series, however, because we feel it provides a strong understanding of some of the newest and most important ways in which urban, regional, national, and international forces are linked to and affect each other. We are convinced that in the present phase of economic globalization, as national, regional, and local economies become tied more directly to the world economy, urban scholars must possess a keen understanding of how the actions of federal states have implications for lower levels of governments and for people in the places in which they live.

States, especially prominent ones such as the United States, are both driving and responding to international pressures to reconstitute their traditional practices and responsibilities in order to participate more flexibly in the global political economy. As these states transform themselves they also force lower levels of government—local states—to do the same. Thus, in order to understand what is happening to and in cities, it is essential to understand how national governments and newly emerging international governing entities are creating new context within which local governments must operate. *State Devolution in America* is an important step in the direction of answering this need in that it provides not only insightful assessments of the impacts of various aspects of federal devolution but also demonstrates how further research into devolution and the restructuring of government might be successfully directed.

David Perry and Sallie Marston
UAAR Series Editors

Acknowledgments

As with all intellectual efforts, compiling this book has been a cooperative effort involving many people and organizations, some of which are not evident in the list of contributors. The initial idea for the book was developed at a workshop sponsored by the Association of American Geographers (AAG) held in July 1995. The workshop was organized by Ron Abler, Janet Kodras, and J. W. Harrington. John Paul Jones III, Robert Lake, Lynn Staeheli, Marvin Waterstone, Jennifer Wolch, and Julian Wolpert also attended. At that workshop, participants organized the Geographers' Network on American Politics and developed a framework for examining state restructuring from an explicitly geographical perspective. That framework is reflected in the introduction to the book and in several contributions. Although not all of the participants at that workshop have contributed to this volume, we are indebted to the group and the AAG for the inspiration and support they provided.

Initial versions of these chapters were papers presented at a series of panel meetings at the 1996 annual meeting of the AAG held in Charlotte, North Carolina. At those panels, the papers and authors received helpful comments from panelists Vera Chouinard, David Hodge, Amy Glasmeier, John Paul Jones III, Laura Pulido, and many members of the audience.

As we began the process of constructing this volume, several other people provided advice, comments, and assistance. We are grateful to Susan Clarke, David Perry, and Catherine Rossbach for their help. Sallie Marston provided guidance, encouragement, and quick reviews when we most needed it. Finally, Rima Wilson and Leigh Miller, at the University of Colorado, provided editorial assistance.

Introduction

LYNN A. STAEHELI,
JANET E. KODRAS,
and COLIN FLINT

The 1994 congressional elections are popularly regarded as a watershed event in American political and social history. Political pundits focus on the election of a Republican majority in both houses, including a large number of activist conservatives, and the rallying of those forces in the first concerted attempt by Congress in over a century to set the course of the country through a dramatic reorganization of the scale and scope of government. The defining document in this effort to restructure government was the House Republicans' *Contract with America*. The Contract and subsequent initiatives advocate an extensive reduction in the size and influence of the national government by devolving federal responsibilities to states and localities, shifting functions to the private sector and civic institutions, dismantling regulations, and in some cases eliminating programs and funding altogether.

Perhaps most importantly, the 1994 election elevated the discussion of state restructuring from a bureaucratic issue—as in "reinventing government"—to political philosophy. The debates catalyzed by the Contract renew fundamental questions concerning what government can and should do, and what level of the federal hierarchy should oversee different functions. The attempt to protect individual rights within a system of collective self-governance and the search for appropriate divisions of responsibility between the national, state, and local levels of that government are ongoing processes within the U.S. federal system—political struggles rooted in the country's Constitution (Sandel, 1996). Seen in the long-term perspective, the current debate is not a rupture with the past, and the 1994 election is not the watershed event that it is so often claimed to be.

At the heart of this ongoing debate is the issue of geographically distributing power and resources within the federal hierarchy in a manner that responds to the diverse map of regional conditions, needs, and preferences, yet protects national norms valued in a common political culture. Current efforts to alter the federal system through devolution, privatization, and dismantling involve a geographic rearrangement of the benefits and burdens conferred by government and thus a redistribution of power and resources among places. Even avowedly aspatial policies administered by the national government vary in their impact and effectiveness when implemented in different jurisdictions and diverse place contexts across the country. Devolving federal responsibilities to states and localities or shifting functions to a highly fragmented private sector greatly accentuates spatial variations in government provision. Although the pieces of legislation enacted subsequent to the 1994 election will have complex and even contradictory effects (Hosansky, 1996), the clear message of the Contract, and the prevailing direction of change, is to shift power from Washington, DC, to the states, thus removing the traditional leveling role of the national government and accentuating differences in the economic competitiveness and social well-being of Americans, depending on where they live. In short, the 1994 elections brought to the forefront of political debate questions related to the role of the government within the larger society.

In a nation as politically fragmented and geographically diverse as the United States, the reciprocal relationship between changes in society and shifts in the form and function of the government is not experienced uniformly in all places across the country. Societal attitudes and sentiments may coalesce at times to produce an apparently single chorus demanding governmental change; we hear this clearly in the current disparaging of the Washington bureaucracy. But some voices carry farther than others, and the individual voices that comprise the chorus differ substantially from place to place, depending on specific conditions in the locale, the ability of individuals to make themselves heard, and the ways in which individuals frame the problems that restructuring is to address. It is thus both useful and important to identify the various constituents of support for governmental change, their economic, political, social, and geographic "locations," and their strategies for change, rather than to assume a single, concerted call to transform societal governance. Furthermore, if these multiple voices are able to effect a change in government, the impact of such changes will differ from place

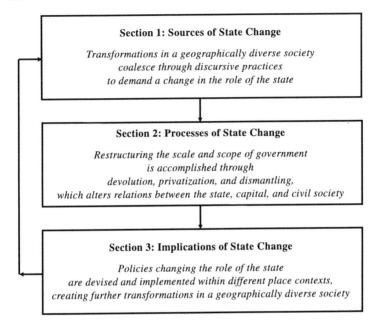

Figure A.1. Linking the Sources, Processes, and Implications of Government Restructuring

to place, depending on the particular local contexts in which they are implemented.

These reciprocal relations between society and state are heuristically portrayed in Figure A.1. Through these relationships, transformations in a geographically diverse American society can generate demands to change the form and functions of government, itself spatially structured and territorially complex, given the nature of the federal system. The resultant government restructuring is instituted across a variegated social landscape, generating place-specific effects in the further transformation of society.

The purpose of this book is to investigate the geographic sources, processes, and implications of the current round of state restructuring in the United States. Chapters in Part I examine recent societal transformations as the sources of pressure for governmental change, specifying their variable character and effect in different local and regional contexts and identifying how these societal changes are transmitted into calls for government restructuring. Chapters in Part II focus on the

Introduction

resulting processes of government restructuring, specifically, how devolution, privatization, and dismantling alter the role of a spatially complex state relative to a geographically diverse society. Part III examines the implications of these current strategies to restructure government in further transforming society, via their effects on specific program arenas such as welfare, immigration, education, industrial labor relations, farm policy, and the environment.

The book is organized to provide an analysis of contemporary governmental restructuring grounded in a geographically differentiated society. To accomplish this goal, the contributing authors draw on the conceptual framework presented in Figure 1, emphasizing three connective themes. First, the contributors address the key question of why state restructuring is occurring—who instigates the demand for change, whose interests are served and whose may be harmed by the changes underway. The answer to this question comes into focus when one discovers how the current process of restructuring shifts the roles and responsibilities of the state in its relation to civil society and the private market, thus rearranging the distribution of power and resources toward particular segments of society and away from others. Second, given that this can be a conflictual process, the authors investigate *how* broad segments of the American populace were convinced that government restructuring was in their own best interests. Here the authors highlight how dominant forms of political discourse—the ways in which issues are framed and discussed—brought together a wide variety of agents in the political, economic, and civic realms to achieve a remarkable, though perhaps temporary, consensus on both the goals of state restructuring and the means of achieving those goals. Finally, the contributors to this volume all work from the perspective that *where* state restructuring occurs is fundamental to understanding the nature and results of the process. The authors demonstrate the importance of spatial scale and place context in creating a desire for state change, in shaping the types of state change proposed, and in generating consequences of state change once enacted. Taken together, these three themes help us to understand the ways in which transformations in the larger society, as experienced in different places, are mobilized to create changes in specific policy arenas and to redefine the role of the state in the American polity.

Although the authors often address specific policies that have been either proposed or enacted, the book should not be considered an institutional analysis of state restructuring. We have refrained from such a project for two reasons. First, as a practical matter, it is too soon to effectively analyze the multiple impacts of current efforts to downsize the government as these play out across the country. Chapters were written on the eve of the 1996 elections when policies emanating from the Contract were in various stages of formulation, and their eventual effects will be years in the making. Second, we are concerned here to demonstrate the value of bringing an explicit geographic perspective into the political debate on government restructuring. In stressing how an analysis of state change is enhanced when studied within the context of a territorially differentiated society, we wish to show how geography complicates not simply this round of restructuring, but others to come. The changes ushered in by the 1994 elections are therefore the beginning, not the end, of our analysis. In the remainder of this chapter, we outline the contents of the book and introduce the three connective themes that bind the individual chapters into the larger message of the volume.

■ Organization of the Book

Part I examines the sources of state restructuring in a society undergoing rapid economic and social transformation (Figure A.1). Robert Lake presents a broad overview of the process by which dislocations in the U.S. economy—occasioned by rising international competitiveness, increased capital mobility, and globalization of production—threatened many segments of American society, such as deindustrialized blue-collar workers, the disaffected middle class, large capital, and small businesses. Although these disparate groups were variably affected by the economic changes, politicians seeking to diminish the role of government were successful in joining these typically contradictory interests into a temporary coalition, persuading sufficient numbers in the electorate that the state, rather than the economy, was the source of their problems.

Remaining chapters in Part I provide a detailed discussion of the issues raised by Lake. John O'Loughlin specifies the ways in which economic transformations have polarized the American population into increasingly isolated affluence groups with variable abilities to maintain a foothold in the dynamic economy. In a parallel piece, Janet Kodras shows that the effect of these economic shifts in splintering the U.S. population has a distinct social dimension, as transformations in the

Introduction

political economy alter not only class relations, but also relations defined by race, gender, generation, and other facets of social life. Both authors illustrate that deepening societal fragmentation, differentially expressed across the American landscape, has created a sense in which the state should be sensitive to region and place-based conditions. They lay the groundwork for understanding why very different societal groups were receptive to ideological calls for state retrenchment. In the final chapter of Part I, Lynn Staeheli addresses the nature of these ideological shifts. She reviews current trends in political rhetoric that emphasize notions such as citizenship and community, and demonstrates the ways in which public attention is being redirected from individual rights to a concern for responsibilities exercised in the context of community and at the local scale. This ideological shift is consistent with the restructuring of government that emphasizes the retrenchment of government programs and regulation, the transfer of public responsibility to the private sector, and the devolution of functions to the state and local scales.

Taken together, the chapters in Part I illustrate how a society as economically, socially, politically, and geographically diverse as that in the United States could join in what appeared to be a single chorus for change in the state. Despite the fact that the specific proposals for state restructuring threatened to further fracture the American polity, many came to believe that, in the political rhetoric of the times, government was the source of the difficulties facing their businesses, families, and communities. At the same time, public faith in capital was declining as corporate profits and stock prices soared, paid for by downsizing and worker lay-offs. Thus, it is understandable that, when faced with a choice of solutions rooted in the state, the private market, or civil society, sufficient majorities of the electorate were persuaded that shifting responsibilities onto civil society was the best alternative. Although polling data suggest that most voters had not heard of the *Contract with America* at the time of the 1994 election, there was, nevertheless, widespread popular support for reducing the scale and scope of the state and expanding the responsibilities of a civil society working in the context of local communities.

Part II examines the processes of current political restructuring (Figure A.1). In her second chapter, Janet Kodras describes the major strategies for state restructuring: devolution, privatization, and dismantling. Each raises questions of spatial scale and place context in the deployment of government functions. Kodras demonstrates that the current strategies will accentuate geographic variations in government provision in accordance with the capacity of local and state governments, the private sector, and social movements in civil society to take on new responsibilities. In the remaining chapters of Part II, Julian Wolpert and Don Mitchell focus on this last issue. Wolpert discusses the capacity of the nonprofit sector to assume responsibilities shed by the state, and Mitchell examines the potential for political mobilization by social movements. Both authors contend that the changes that have been presented as reinvigorating civil society—an increased emphasis on communities to meet the needs of the public, to contextualize rights, and to ensure that individuals fulfill their social responsibilities—may have the opposite effect. Wolpert demonstrates that the charitable sector is unprepared to absorb the burden of meeting social service needs, particularly in those places where needs are greatest. Mitchell argues that the ability to mobilize social movements for political change requires a strong commitment by the federal government to safeguard the notion of individual rights, which is problematic, given the strong ideological attack currently leveled against the concept of rights.

Part III addresses the implications of current state restructuring strategies within specific policy arenas (Figure A.1). It begins with a discussion of policy changes in the major economic sectors. Brian Page examines recent alterations in U.S. agricultural policy as manifested in the 1996 Farm Act. He finds that the changes enacted for this sector—expanding the government's role to assist the global competitiveness of farm producers while retracting its traditional responsibilities to safeguard the family farm and preserve rural life—reflect general political currents that prioritize demands made by powerful players in the U.S. economy. His review of the battles fought over design and passage of the Farm Act reveals how large-scale producers pressured the state for change with little effective resistance from small producers. The effects of these policy shifts will differ substantially by scale of production, commodity type, and region, granting favor to agribusiness and to specific commodity producers, especially those operating in the South and West. Andrew Herod finds the same set of priorities in current attempts to rewrite labor law and employment regulations in the industrial and service sectors: a preference for helping large-scale producers to operate on the global stage, while revoking national safeguards that had protected U.S. workers. Proposals made by both the Republican-majority Congress and Clinton's Dunlop Commission would have the effect of globalizing capital and localizing labor, placing workers at a distinct negotiating

Introduction xix

disadvantage. The result would be an increasingly variegated landscape of labor relations shaped by the heightened power of producers.

The remaining four chapters in Part III address policies affecting or feeding into these employment sectors. Meghan Cope examines the issue of welfare reform, which began well before the passage of the 1996 Bipartisan Welfare Reform Act. She argues that as international economic transformations have heightened competitive pressures on U.S. firms, capital has demanded retraction of the welfare state that, it argued, priced U.S. labor out of the global market and provided disincentives to work at the lower end of the labor market. The Welfare Reform Act accomplishes this goal by ending the federal guarantee of support for poor families and devolving responsibility to state and local jurisdictions, where "competitive downsizing" will generate an increasingly uneven policy terrain of assistance programs. Cope shows that the *spatial* strategy of devolution was a key component of the *political* strategy to end universal support for the poor, as the effects of this policy shift are concealed in the thousands of programs implemented in local jurisdictions across the country. Societal support for ending the welfare state was purchased with an assault of political rhetoric that convinced middle-class voters that the welfare state had diminished personal responsibility among the poor and that strong incentives, such as workfare, were needed to correct the problem.

Richard Wright addresses the impact of state restructuring for another disadvantaged group in U.S. society, examining the plethora of policy changes affecting legal and illegal immigrants. He identifies a great contradiction in the role of the state, which simultaneously seeks to halt the global flow of labor—seen to threaten the job security of Americans—even as it supports the global flow of capital through free-trade policies and the like. Attempts to blame immigrants for the deteriorating economic conditions of working-class Americans and the breakdown of "American society" deflect attention from the role of capital seeking profit in the intensifying global competition. The resulting nationalist attitudes feed into current efforts to restore culture, values, and economic control to a uniquely American community that are part of the larger discursive goal of state restructuring. Both Cope and Wright pay particular attention to the ways in which the effects of state restructuring in these policy areas are cross-cut by gender, race, nationality, generation, and region. They argue that the result of restructuring will be to further fragment geographic patterns of well-being and heighten inequality across the country.

The final two chapters in Part III show how the complex and inconsistent roles of the state are thrown into sharp relief during the process of government restructuring. Fred Shelley studies the contradictory position of the state in education policy as it attempts to meet the labor requirements of capital, which are national in scope and would benefit from greater coordination, yet to also address parental desire for local control over their children's education. As a result, the general political trend toward devolution and dismantling of federal control is much more problematic in education policy. Although there is some pressure in this direction, Shelley argues that social conservatives fear the outcome if local school districts are given complete control over curricula. Demonstrating the importance of place, Shelley notes that inner city schools might take advantage of such freedom to promote multicultural education, rather than the "traditional American values" promoted by social conservatives.

Marv Waterstone examines recent battles over environmental policy, a clear target of congressional conservatives although it was never specifically mentioned in the Contract. Through his review of the myriad legislative proposals addressing water pollution, wildlife preservation, public health standards, natural resources, and the like, Waterstone demonstrates that in this particular policy realm, efforts to diminish the role of the national government through devolution, privatization, and dismantling have often met effective resistance by environmental movements marshaling strong public sentiment. In reaction, opponents of environmental regulation have pursued changes through the labyrinthine budgetary process, cutting funding rather than programs. Because it is much more difficult to mount public resistance to hidden budgetary maneuvers than to high-visibility programs, the opponents of environmental regulation have been more successful in pursuing this strategy, and the long-term effects of restructuring environmental policy will be relatively difficult to assess.

In the conclusion of the book, Colin Flint accumulates the messages from the empirical chapters and revisits the theoretical issues raised by Lake. He argues that through actions taken at a variety of scales and places, and by many different actors, state restructuring primarily strengthens the position of capital, even though the changes are presented as supporting civil society. As capital gains global competitive muscle through such policy shifts, the power of the national state diminishes, throwing subnational regions and localities directly into competition within the global economy and bolstering the political significance of "place."

Powerful regions are best positioned economically and politically to influence federal policy by imposing their own geographically specific and historically accumulated traditions and ideologies on the federal government, thus restructuring the national perspective on the nature and role of the state. The South, in particular, has been successful in advancing onto the national agenda a regional ideology promoting particular priorities and values, such as personal responsibility, private property rights, and "traditional family values." Furthermore, the South's ideological predilection favors a particular stance toward the government, employing the rhetoric of governmental nonintervention and supporting states' rights over the power of the national government. One of the key ideas to emerge from this analysis is the understanding that state restructuring in a territorially diverse society is not solely "top-down." Ideologies and material conditions rooted in place can be mobilized in national-level debates to suggest strategies for restructuring the state and to present an idealized model of social relations for the country as a whole.

■ Connective Themes

Having presented an outline of the book, we now describe three predominant themes woven throughout the individual chapters and binding each to the conceptual framework presented in Figure A.1.

Restructuring the Relations Between the State, Capital, and Civil Society

Studying the changing role of government requires a particular conceptualization of the state. We take a pragmatic view of the state as a set of institutions and agents that interact with institutions and agents outside the state. Thus, the state is neither completely autonomous of, nor completely dependent on, the wider society of which it is a part (Clark & Dear, 1984; Jessop, 1982, 1990). Indeed, the authors in this volume seek to understand the process of government restructuring by examining the shifting relationship between three principal spheres of American life: *the state, capital,* and *civil society*. We focus on the state not to privilege it, but to understand the ways in which these spheres are mutually constituted through their interactions.

Both the motivation for, and the effects of, state restructuring—who instigates the demand for change, who stands to benefit and who to lose—become clear when one determines how the process of government restructuring shifts relations between the three spheres, thus redistributing power and resources toward particular segments of society and away from others. The key is to ascertain how the primary roles of the state are altered through restructuring to rearrange the benefits and burdens conferred by government. Following other theorists of the state, Lake argues that two state roles are paramount: *accumulation* and *legitimation*. In the simplest terms, one function of government is to provide a set of conditions and resources to promote economic growth and the accumulation of capital. The second role of the state in a liberal democracy is to seek legitimacy from civil society. To accomplish this goal, the state must maintain the faith and loyalty of citizens by protecting individual rights and representing the "will of the people." The state is under constant pressure to balance these two, often contradictory roles in ways that simultaneously bolster the power of capital and protect the interests of citizens. Seen in this light, government restructuring is a process whereby the state renegotiates its dual role with capital and civil society, a politically contentious recalibration with important consequences for who bears the costs and who obtains the benefits of government resources.

Before we summarize the insights gained by applying this approach to the present round of policy changes, a few initial points are in order. First, the state, capital, and civil society are complex and overlapping entities. The *state* is particularly complicated, consisting of disparate agents and institutions, working at different levels of the federal hierarchy and in distinct places, pursuing different and often inconsistent objectives. As Mitchell reminds us in his chapter, the state can be either society's strongest guarantor of human rights, or it can itself threaten and kill. Similarly, *capital* encompasses a wide range of often contradictory organizations (big vs. small business, transnational corporations vs. locally-dependent firms) and sectors (agriculture, heavy manufacturing, high-tech manufacturing, services). As Lake describes in his chapter, these various entities have weathered globalization of the U.S. economy in different ways, affecting the interests each holds in particular proposals for a governmental response to economic change. More specifically, Page illustrates how the 1996 Farm Act portends different effects for various facets of the agricultural sector according to scale of production, commodity type, and region. *Civil society* is also diverse and is currently fragmenting into increasingly isolated segments as recent transformations in the U.S. economy have altered social struc-

Introduction xxiii

tures and relations. O'Loughlin describes how economic restructuring has repositioned different societal groups to increase income inequality in the United States. Taking the argument a step further, Kodras demonstrates in her first chapter how this economic polarization is further fragmenting the population along the lines of race, gender, and generation to redraw the geographies of inclusion and exclusion in American society. Wright presents a particularly subtle and interesting example in describing the new complexities of nationalist identity and sentiment that feed into debates over immigration policy.

A second point to bear in mind is that the boundaries between these three complex spheres are porous and dynamic. One example of the overlap between the state and capital is the public-private partnership established in many cities and states to promote economic development. The indistinct boundary between the state and civil society is evident in the example of nonprofit organizations providing government-funded services. And an example of the blurred margin between capital and civil society is seen in the fact that agents operating in the economic sphere are also individuals living in the civil society; they have families and are of different ages, races, classes, and genders, all of which define their interests in civil society. The same economic agents who applaud a state action favoring accumulation may question the "rightness" of that action in its social effects.

A third point worth remembering is that the accumulation and legitimation roles of the state are not mutually exclusive, as evidenced by polls that document declining confidence in governmental institutions during times of economic hardship (e.g., Greenberg, 1995). When the material conditions of large numbers of households stagnate or decline—as O'Loughlin and Kodras demonstrate is occurring—citizens can look to the state, capital, or civil society for recourse. Fainstein (1987) argues that Americans are ideologically predisposed to blame government, and as a result, the state is often placed in the inconsistent position of counteracting the loss of legitimacy that may follow from its own accumulation policies. These three points make clear that while the language we use must necessarily simplify and isolate the state, capital, and civil society and also differentiate between the accumulation and legitimation functions of the state, these spheres and roles are in reality more complicated and entangled.

Each of the chapters in Part III reveals how the current round of governmental restructuring is actively renegotiating the twin roles of the state, with dissimilar consequences for different segments of American society. Page shows how the traditional dual role of U.S. agricultural policy, bolstering profitability through a system of subsidies and price guarantees while seeking to preserve the family farm and protect rural traditions, has been fundamentally altered by passage of the 1996 Farm Act. Large-scale producers operating in the international economy have pushed for changes in agricultural policy to help their competitive position. In the process, the interests of globalized agribusiness and locally dependent small producers in sustaining rural American life have diverged. Because the former hold greater power to influence state policy change, the 1996 Farm Act has the effect of strengthening the accumulation function, to the detriment of small producers and of the economic health of rural America. Significantly, legitimation concerns expressed during congressional debate leading up to the bill's enactment focused on protecting the environment, where opponents found strong public support, rather than on saving the mix of credit and commodity programs that had served as a social safety net for small farmers.

Herod examines workplace regulation, whose traditional dual function it has been to explicitly negotiate between capital and labor, furthering the growth and profitability of industrial firms while protecting workers' health and safety and ensuring just compensation. Current policy recommendations are primarily driven by accumulation concerns, with both Republican and Democratic groups arguing the need to assist the international competitiveness of U.S. firms. Herod forces us to consider the effects of these proposals on the legitimation function of the state, which, he argues, will further localize and subordinate labor to the needs and demands of capital. It is in the state's dealing with labor as simply an economic group—conceptually stripping workers of their status as family members, as people with diverse backgrounds, interests, and characteristics—that the attempts to redefine the boundaries between the state, capital, and civil society are most clear.

In their respective chapters on welfare and immigration policy, Cope and Wright find that the current set of policy changes will further impoverish and marginalize poor and foreign populations. Welfare policy has conventionally served as both a regulatory mechanism, maintaining a potential workforce as needed by the shifting demands of capital, and a safety net, protecting workers against the vicissitudes of the market. But in an era of accelerating international competition, capital has sought to reduce labor costs by, among other tactics, attacking the welfare state and thus ensuring a labor force willing to work in contingent, low-wage, and low-benefit jobs. The result is the 1996 Bipartisan

Welfare Reform Act, an extensive devolution, privatization, and dismantling of the previous system, leaving little alternative for labor but to accept the punitive conditions set by capital. Wright demonstrates especially strong contradictions in U.S. trade and immigration policies, as the state simultaneously supports the forces of transnationalism and nationalism. Specifically, the state promotes the global accumulation of capital through neoliberal, free-trade policies, but as public resistance to globalization increases, the state seeks legitimation from civil society, enacting exclusionary immigration policies that appear to lessen domestic job competition and welfare reform that denies benefits to noncitizens.

The final two chapters in Part III highlight the point that state restructuring is a highly conflictual process of negotiation between different segments of American society, with no ensured winners. As Waterstone details in his chapter, the state's contradictory accumulation and legitimation functions with regard to the environment generated a particularly heated conflict when Congress considered revoking key regulatory programs. Environmental groups mobilized sufficient public resistance to stop many of the original proposals from being enacted into law. The battle continues, however, as opponents of environmental regulation have since taken the subterranean route of weakening extant laws through budget cuts. Finally, Shelley demonstrates that the processes of state restructuring may generate perplexing internal conflict when a particular group becomes entangled within the contradictory roles of the state. He notes the difficulty that education policy represents to those concerned with the state's teaching of values. While social conservatives want fewer restrictions on how values are taught in the schools, they are also concerned about the particular kinds of values that might be promoted if all regulations were dismantled.

In summary, the authors in this volume reach the general conclusion that contemporary processes of governmental restructuring are having a demonstrable effect in redrawing the boundaries between the state, capital, and civil society, and that this is being accomplished in ways that tend to favor large capital. The result has been to shift further the burden of ensuring adequate living conditions from capital and the state onto civil society. It is in this context that Wolpert's assessment of the charitable sector—that element of civil society presented as mitigating the effects of economic transformation and political restructuring—is so important. Wolpert questions the ability of the charitable sector to assume added responsibility, particularly in the face of rising income inequality and the differential capacity of local places to respond to the needs of area residents. In short, he finds little evidence that those who have benefited from economic and political changes are willing to fund, through the charitable sector, the kinds of services that ameliorate problems, and he argues that those places most in need are the least able to respond through charitable giving. Mitchell then addresses the critical question of how civil society can effectively respond to the current processes of governmental restructuring. He argues that social movements must engage the state, seeking to draw it back in line with their interests by exploiting the state's traditional obligation to represent "the will of the people." Stated in another way, civic groups opposing the state's current focus on its accumulation role need to reassert the legitimation function of government.

Given the argument that governmental restructuring has generally been accomplished in favor of capital, how is it that "average Americans" came to believe that giving more power to this sphere was in their interests? How is it that working-class Democrats came to believe that reducing government regulation was to their benefit? How is it that African American mothers living in central cities were held responsible for rising welfare costs and declining moral standards? How is it that immigrants were made accountable for the breakdown of American culture? The authors in this volume look to the role of political discourse in state restructuring, the second connective theme in our conceptual framework, for their answers to these questions.

The Discourse of State Restructuring

Lake argues that the answers to the questions above lie in the ability of political cadres in both the Republican and Democratic parties to mobilize political opinion in ways that made the interests of a wide variety of economic, social, and political groups appear to converge. Their ability to do so rested on a discourse that framed both the problems state restructuring should address and the solutions to those problems. In this discourse, inequality was minimized, the plight of the middle class was linked with that of capital, and responsibility for the well-being of individuals and families was shifted onto civil society.

Staeheli addresses the latter element of the discourse of state restructuring. She notes an increased emphasis on personal responsibility for solving problems relating to families and on community as the setting in which personal responsibility can be fostered. The discourse of responsibility dovetails with the three forms of state restructuring. The combined effect of devolution, privatization, and dismantling federal programs is said to provide the conditions wherein citizens can reconnect with community in self-government and develop the virtues of a responsible citizenry. The key to addressing the dramatic changes in the social and economic structure, then, is held to be a revitalized citizenry that draws resources from civil society rather than from the state.

The discourse promoting responsibility, virtue, and community struck a resonant chord with white, working-class voters anxious about their future. As Stanley Greenberg (1995) notes, these "downscale Democrats" were the swing votes that propelled the 1994 Republican victories in the U.S. Congress. Only vaguely aware of the *Contract with America*, this group did know that its own future looked bleak. Increasingly dissatisfied with a tax burden made more oppressive as their economic position eroded, they were receptive to arguments that the sources of their problems lay in a government far removed from everyday people and in social groups who received benefits outweighing their contribution to society.

The solution for many people thus appeared to be a withdrawal of the state from both capital and civil society. While many citizens responded to the call for community and responsibility in civil society—for the redrawing of boundaries between the state and civil society—less attention was given to the effects of restructuring on the boundaries between the state and capital. To the extent that attention was directed to the latter, it tended to reinforce the American ideology that the market will provide solutions to worsening material conditions if only the government would get out of the way. Thus as Herod notes, there was little public reaction when the repeal of state policy ensuring basic labor negotiating principles was justified as necessary to help U.S. industrial firms flexibly compete in the world economy. And as Page points out, few in American society saw the discursive irony of giving the 1996 farm policy changes the label, Federal Agricultural Improvement and Reform (FAIR) Act, although the policy shifts were hardly an equitable arrangement, promoting the international competitiveness of agribusiness while removing traditional safeguards that had protected small producers and sustained rural communities.

The prevailing discourse surrounding state restructuring thus deflected attention from crises in the global and national economies that created hardship for Americans, focusing instead on a government out of touch with common folk and on the unfair advantage given to "less deserving" people. With regard to the latter, recent political rhetoric has effectively identified scapegoats—segments of society who can be made to take the blame for worsening economic conditions. It is in this context that Wright's chapter on immigration policy and Cope's chapter on welfare reform are informative. Listening to the rhetoric of recent election seasons, one could reasonably conclude that immigrants and welfare recipients are responsible for the federal budget deficit, the economic distress felt by the "average American," and the deterioration of cultural values. It is not surprising, then, that the most notable "successes" in restructuring the government have been the passage of welfare reforms and restrictions on immigrants. Health care reform, which would likely do more to alleviate the effects of accumulation policies, foundered in Congress due to opposition from capital. Only Patrick Buchanan broached the problem of rising income inequality in the 1994 election campaigns, but he also tended to blame such problems in large part on immigrants and welfare recipients.

The ability to present problems that arise from shifting economic conditions as being the fault of changes in civil society demonstrates the porousness of the boundaries between capital and civil society and the ability of politicians to discursively manipulate those boundaries. Mitchell calls on social movements in civic society to recognize the political purposes behind the prevailing rhetoric and to devise their own discursive strategies as a counterforce. Waterstone demonstrates the power and potential of using this approach in his examination of recent battles between environmental groups and congressional opponents of environmental regulation. Mitchell argues further that the ability of marginalized social groups to shape public opinion and to preserve the rights of individuals is threatened by the current discursive emphasis on responsibility and community. The implications of such a shift in balance between rights and responsibilities and between individuals and community are profound, speaking to the very nature of liberal democracy in the United States.

The Role of Space and Scale in Political Restructuring

To this point, we have focused on the ways in which the discourse of state restructuring has redefined relations between the state, capital, and civil society. Occasionally, spatial metaphors such as "boundaries," "maps," and so forth have been used. But the new geographies created by state restructuring are material, as well as metaphorical. The contributing authors in this volume work from the perspective that *where* state restructuring occurs is fundamental to understanding the nature and results of the process. The authors demonstrate the importance of spatial scale and place context in creating a desire for state change, in shaping the types of state change proposed, and in generating consequences of state change once enacted. The current debates over state restructuring can thus be seen as part of an ongoing struggle to create new geographies in the United States through the manipulation of scale and space.

This argument is consistent with what geographers call a "social constructivist" perspective, whereby the spatial organization of the United States—the spaces and places in which we live and work—are produced through the actions taken in the spheres of capital, the state, and civil society. Of relevance here, the structural conditions underlying the call for governmental restructuring are experienced in different ways throughout the country, and the resulting changes in the state create new geographies in the further transformation of American society.

Spatial scale refers to the level at which processes operate in the world system, from the global to the local levels. Agents and institutions of the state, capital, and civil society all work at these various levels. The role of spatial scale is most explicitly discerned in processes operating through the state, given the distinct national, state, and local tiers of the federal hierarchy. As Flint notes, one of the central features in recent debates over state restructuring is to figure out the scale at which decision making, service delivery, and accountability for government action is most appropriately located. But scale features prominently in the operations of capital and civil society as well. In Part I, Lake, O'Loughlin, and Kodras describe the general processes whereby dislocations created as local economies reposition within the national economy, itself repositioning within the dynamic global economy, create a varied geographic pattern of stress and hardship for different segments of capital and for different segments of the American population. These disjunctures and discontents play out in the political arena, as agents and institutions representing capital and civil society engage the state, pressing for governmental changes that best secure their own position. Here they confront a particular set of political opportunities available at each tier of the federal system. The ability of a particular group to gain access to the state at a given tier changes over time; Kodras explores in her second chapter the ways in which political opportunities are shifted between levels of government through the process of state restructuring. Wolpert and Mitchell then discuss the ways in which these changes in the state influence the structure of political opportunities available to the charitable sector and social movements in civil society.

At any given time, agents and institutions attempting to secure position through state restructuring work at the level of government presenting the greatest political opportunities for their purposes. They may also attempt to control and strengthen the particular scales of government that benefit their agenda and to weaken the levels at which they face opposition. In their respective chapters, Mitchell, Herod, and Kodras demonstrate that local states, social movements, and civic groups can gain power by forging cross-scale alliances, networks extending upward to kindred groups operating at different scales and reaching outward to similar groups in other places. These alliances draw in external resources and expertise, expanding the power of local groups to participate in the political process. The three authors cite examples of scale-jumping tactics and alliances featured in recent political skirmishes, such as the Colorado gay rights initiative.

Place context is a second geographic dimension shaping the sources, processes, and consequences of state restructuring. Local places develop distinctive economic characteristics (based on local resource endowments, production systems, linkages to the national and global economies, etc.), political practices (grounded in factional ideologies, political party dominance, civic and philanthropic traditions, etc.), and social relations (defined by class, race, gender, generation, etc.). As Kodras describes in her second chapter, these conditions combine to create particular institutional contexts—accretions of organizational experience and political seasoning, monetary resources, and infrastructure—that shape the capacity of governmental and nongovernmental agents and institutions in each place, and indeed, the local sense of what is possible and appropriate for these groups to undertake. As a result, state and local governments, private corporations, nonprofit organizations, civic groups, and individuals situated in different places vary considerably in their ability to generate the demand for state restructuring and in their capacity to respond to the changes thus enacted.

Unfortunately, most U.S. policy is insensitive to place; and most political analyses overlook the role of place in molding the needs, interests, and power of political agents. The widespread failure of policymakers to consider how the effectiveness of a program depends on the local context into which it is inserted undermines even the best of intentions. This point holds both for national-level policies imple-

Introduction

mented at lower tiers of the federal hierarchy and for local initiatives that are diffused to other places without due consideration for the very different circumstances to which they are applied.

To help correct this oversight, authors in this volume analyze the effect of place context in shaping the outcomes of specific policy initiatives. Of particular concern is the question of capacity: Which states have the ability to develop and administer new welfare provisions? Which localities have a charitable sector that can absorb the loss of social service funding? Which states have weaker regulatory and enforcement capabilities? And ultimately, how does a geographically diverse policy terrain and set of institutional structures shape the geography of inequality in American society?

The authors in Part III address the issues of place context and capacity to demonstrate how specific policy changes could further fracture the nation along geographic lines. They reach the common conclusion that state restructuring will increase current disparities in the economic and social well-being of the American population, making it more difficult for citizens to join in struggles that might counteract the effects of this round of state restructuring. Many of the contributors draw on specific place examples to illustrate how general processes of state restructuring are apt to play out geographically. Notable examples include Page's prognosis of the variable effects that the 1996 Farm Act will have on the Southern, Western, and Midwestern commodity-production regions; Cope's assessment of the impact the 1996 Bipartisan Welfare Reform Act will have on places of entrenched poverty, such as the Mississippi Delta; Lake's analysis of recent trends in New Jersey as emblematic of the economic transformations fomenting demands for state restructuring; and Mitchell's description of a prolonged labor strike in Decatur, Illinois, to demonstrate the difficulties social movements have in organizing effective resistance against capital. In the concluding chapter, Flint provides the most detailed example of the role played by place context in current state restructuring, showing how the South advances a regional ideology of the state that resonates with the dissatisfactions many Americans feel toward their government. Employing specific examples such as these to illustrate how general processes come to ground in particular place contexts is a hallmark of geographic research, countering a pervasive tendency in many of the social sciences to cite "facts from nowhere" and to represent processes as occurring "on the head of a pin."

■ Summary

The underlying argument presented in this book is that governmental restructuring involves a rearrangement in the relations of the state, capital, and civil society, with the outcome depending on the needs and goals of the state itself and the ability of agents and institutions in all three spheres to shape political discourse and policy direction. It is in this sense that the *Contract with America* is emblematic of larger forces reshaping the American polity, much as the New Deal was a manifestation of deeper political currents in the 1930s. Whereas the Great Depression motivated calls for governmental restructuring that expanded the role of the state, current economic crises have prompted demands to retract the scale and scope of the state through devolution, privatization, and dismantling. Each of these strategies raises questions of spatial scale and place context—for example, what level of government is best able to address an issue? When should the private sector be responsible for a function? Do local governments and nonprofits have the capacity to take on new responsibilities? How do local social movements find the leverage to negotiate with global capital through the state? The answers to these questions are complex, depending on the material conditions and discursive practices found at different scales of the federal system and in different places and regions across the country. In sketching the overall format and themes of the book, this introductory chapter has provided only the barest suggestion of the fascinating, and often troubling, assessments of state restructuring in this book.

Although the chapters share a common conceptual framework, we want to caution that they do not share a common vision of the specifics of state restructuring or whether the eventual outcomes will be positive or negative. We recognize that in assessing issues as politically contentious as government restructuring, differences of opinion will inevitably arise among informed observers. Because the conceptual framework informing this analysis of state change portrays government restructuring in terms of a shifting relationship between the state, capital, and civil society, it necessarily draws attention to the tensions and conflicts between these three spheres. These stresses have generated new problems for some, yet opened fresh opportunities for others. In addressing their respective topics, the individual authors have attempted to identify both the negative and positive consequences of current state changes for different spheres and segments of American society located in different places. As editors, we have chosen not to censor or alter the assessments and evaluations made by the authors; the authors' views are their own. As such, it is likely that the reader will take issue with conclusions reached by at least some of the authors—just as have the editors. We believe such disagreements reflect the complexity and contentiousness of state restructuring in a diverse society as observed from diverse viewpoints.

The process of state change is an ongoing response to broad dynamics at the global and regional scales. The present round of restructuring seems to be redrawing the boundaries between the state, capital, and civil society in ways that favor capital. But the chapters in the book also remind us that we should be asking not simply, What is happening now? but also, What will happen in the future? Our intent in this book is to provide an answer to the first question, but more importantly, to provide a template of, or basis for understanding, the sources and effects of state restructurings yet to come. A final but unanswered question in the book is whether movements in civil society can be made sufficiently strong to draw the state into a more balanced position between capital and civil society. Given the increasing fragmentation between segments of society and the increasing geographic variation between places that we argue will result from the current round of state restructuring, mobilizing the American population to redraw the boundaries becomes an even more formidable task. Such an effort requires citizens to consciously perceive their particular position within the complex, interlocking structures of the state, capital, and civil society; to understand how discursive practices frame what is possible; and to become actively engaged in politics broadly taken.

REFERENCES

Clark, G., & Dear, M. (1984). *State apparatus: Structures and languages of legitimacy.* Boston: Allen and Unwin.
Fainstein, S. (1987). Local mobilization and economic discontent. In M. P. Smith & J. Feagin (Eds.), *The capitalist city: Global restructuring and community politics* (pp. 323-342). Oxford, UK: Basil Blackwell.
Greenberg, S. (1995). After the Republican surge. *American Prospect, 23,* 66-76.
Hosansky, D. (1996, November 2). GOP confounds expectations, expands federal authority. *Congressional Quarterly Weekly Report,* 3117-3122.
Jessop, B. (1982). *The capitalist state.* Oxford, UK: Robertson.
Jessop, B. (1990). *State theory: Putting capitalist states in their place.* Cambridge, MA: Polity.
Sandel, M. (1996). *Democracy and its discontent.* Cambridge, MA: Belknap.

Part I

Sources of State Change

1

State Restructuring, Political Opportunism, and Capital Mobility[1]

ROBERT W. LAKE

In the turbulent politics of the 1990s, the major political parties agreed on perhaps only one thing: that the era of "big government" was over. But despite the similarity of their rhetoric, this meant quite different things for Democrats and Republicans. For the Democrats, the end of big government meant a hiatus in the establishment of new programs. The rhetoric of government restructuring was their apology for fiscal impotence. It expressed a pragmatic recognition that, despite pressing "social problems"—the name we give to economic problems that affect the poor—the lack of funds resulting from the federal budget deficit precluded the launching of large and expensive new government programs. Under the Clinton administration, the potentially activist and socially interventionary cabinet departments—Housing and Urban Development, Health and Human Services, Labor—remained conspicuously inconspicuous. Whatever the problem was, government could not be the solution.

For the Republicans, government was the problem. The assault on big government provided a rationale for eliminating *existing* government programs and regulations. The Republicans' anti-big-government rhetoric was their answer to the impotence of market economics. It represented a pragmatic recognition that the threat of declining corporate profits required the elimination of government programs and regulations that hindered the movement of capital into more profitable investments. The Republicans' intention, unambiguously announced in their *Contract with America,* was to "roll back government regulations

... enhance economic liberty ... and [break] down unnecessary barriers to entry created by regulations, statutes, and judicial decisions" (Gillespie & Schellhas, 1994, pp. 126-128). They substantially accomplished this goal after 1994 through their control of congressional appropriations and funding bills that slashed agency budgets and explicitly eliminated or curtailed an astonishing array of regulatory activities (see, for example, Cushman, 1995, 1996; Natural Resources Defense Council, 1996). This attack was highly selective. As documented elsewhere in this volume, programs that favored business interests expanded under the Republican-controlled Congress, while environmental, occupational, and workplace safety and health regulations that protected workers were drastically cut (for details, see the chapters in this volume by Waterstone on environmental policy, Page on agriculture, Cope on welfare, and Herod on labor unions).

The severity of these actions and the intensity of the political debates surrounding them raise several enticing questions. Why were Republicans and Democrats alike so intent to declare the end of big government in the early 1990s? What accounts for the vastly different meanings of government restructuring encompassed in the two parties' visions? What explains the ability of the Republican rhetoric of deregulation to combine usually disparate and contradictory political factions to a sufficient extent in 1994 to give that party control of both houses of Congress for the first time in more than forty years? What conditions prompted a political backlash to emerge as the effects of the Republicans' restructuring program began to become apparent, primarily in the environmental, health, and social policy arenas?

Beyond the immediacy of explaining the outcomes of particular national elections, the recent politics of government restructuring also raises some more general questions about the sources of state policy. Delving beyond the rhetoric, what purpose is served by the radical decentralization, privatization, and deregulation encompassed in restructuring the state? Whose interests does it represent? The 1992 election bumper sticker asserting that "It's the economy, stupid" only begs the questions of how and why economic conditions affect the form, functions, and purposes of the state. Why, and in what ways, are changes in the organization chart of the state linked to changes in local, national, and global economies? How are economic needs translated into demands for changing state forms? Who does the translating? Under what conditions? On whose behalf? These questions, in turn, raise even more fundamental and abstract issues pertaining to the relationship between politics, the economy, and the state; the meaning of relative state autonomy; the relative importance of structure and agency in reproducing state formations; and the role of ideology and rhetoric in manipulating the consciousness of the electorate toward politically instrumental purposes with real and important consequences for control of political power and the distribution of state resources.

To explore some of these questions in the following pages, I begin with a brief overview of the idea of relative state autonomy as a way to conceptualize the links between government restructuring and changes in the economic conditions affecting various electoral constituencies. This overview provides a point of departure for examining the dynamic of state restructuring as a response to the crisis in the market economy of late 20th-century capitalism. In the following section, I develop an argument that attributes the Republican Party's success in winning control of Congress in 1994 to its ability to assemble an apparent, if temporary, alliance of usually disparate economic and class interests behind a rhetoric of massive deregulation and government restructuring. Although affected in different ways by recent manifestations of economic crisis, important factions of both capital and labor were temporarily, yet sufficiently, mobilized by the Republicans' rhetorical appeals to support a common solution involving government restructuring. This congruence of interests produced a temporary coalition that House Republicans used to further their short-term partisan purpose of expanding their grasp on political power. Their inability to sustain that coalition in 1996 accounts for their difficulty in achieving further restructuring and expanding their political position.

The framework I explore in this paper, therefore, understands the state restructuring unfolding in the early 1990s as resulting from two intersecting influences. The first is the underlying crisis conditions in the market economy of late capitalism, conditions that made a number of usually disparate interests susceptible to the attraction of simple solutions. Restructuring the state appears as a relatively simple solution to economic crisis in the sense that it is within the power of state actors to achieve, and in the further sense that it leaves fundamental social relations intact. The second influence on state restructuring is the ability of contending political cadres to mobilize the effects of economic crisis conditions for their short-term political advantage. In this view, structure and agency are mutually constitutive of state outcomes through a process in which political actors discursively manipulate the meaning of structural crisis to their political advantage.

Relative State Autonomy

A search for the causes or sources of government restructuring immediately raises the question of the relative autonomy of the state from the economy (Barrow, 1993; Carnoy, 1984). Do the form and functions of the state change in lockstep with changing economic conditions (a state dependency argument), is the state completely autonomous from economic conditions, or is there some intermediate relationship of relative state autonomy?

To examine this question, I begin with the theory of state derivation proposed by Joachim Hirsch (1978, 1981). "State interventions," according to Hirsch, "are essentially determined by the crisis-laden character of capitalist society" (Hirsch, 1981, p. 593). In Hirsch's view, the strategies available to capital to resolve recurrent crises of profitability, such as technological innovations, improved economies of scale, or new product development, are at best temporary stopgaps. The inadequacy of such measures to resolve the inherent contradictions of the market economy drives capital to adopt increasingly violent and drastic solutions: the de-skilling of large segments of the labor force; the destruction of societies and cultures through the commodification of new regions within the sphere of capitalist production; the marginalization, exclusion, and abandonment of other regions; the exploitation and destruction of physical resources and ecological systems, and so on. But these solutions—"the accelerating tendency to violent restructuration ... of every sphere of society" (Hirsch, 1981, p. 593)—are so extreme that capital cannot employ them unilaterally. Instead, capital requires the "intercession of the State" in order both to facilitate their imposition and simultaneously to cushion the resulting social upheaval and conflict. The state, in short, arises in a particular form and adopts specific functions because it is required to do so by capital unable unilaterally to avoid recurrent crisis and ensure its own reproduction.

In this capacity, the state engages in two necessary but contradictory functions: providing the conditions for capital accumulation and, simultaneously, maintaining the legitimation of the social and economic system (Habermas, 1975; Kodras, this volume). In order to do so, the state organizes itself as an input-output mechanism (Habermas, 1973, 1975; see also Cypher, 1980; Solo, 1977, 1978; Wolfe, 1977). That is, the state distributes outputs to capital in the form of the conditions for accumulation (e.g., building infrastructure, maintaining a stable market and financial system, guaranteeing property rights, enforcing contracts, etc.) and distributes outputs to labor in the form of social welfare investments and expenditures (e.g., unemployment insurance, education, welfare payments, etc.). In exchange for these outputs, the state receives inputs from capital in the form of financial resources and receives inputs from labor in the form of mass loyalty (Clark & Dear, 1984). The institutional design of the state apparatus is organized into the particular elements—executive units, administrative departments and agencies, courts, enforcement agencies, etc.—necessary to collect the flows of inputs and distribute the outputs.

Politics in this framework becomes a process for defining, debating, and calibrating the substantive form and relative magnitudes of these inputs and outputs in specific places and moments—that is, a process for deciding who gets (and pays for) what, where, and how. This politics is highly contentious and enormously consequential for who bears the cost and receives the benefits of government resources. Given finite resources, pressure from capital for increased state outputs necessarily reduces state outputs to labor, risking a legitimation crisis through the withholding of mass loyalty (witness the recent vulnerability of incumbents in political office). Conversely, pressure from labor for increased state outputs necessarily reduces outputs to capital, risking a rationality crisis through the withholding of financial resources (witness the export of capital to offshore locations). Most often, political debate focuses on the magnitude of inputs and outputs within existing structures of government (e.g., debates over tax rates and incidence, subsidy levels, regulatory standards, highway expenditures, etc.). Under certain conditions, however, political debate escalates to demand changes in the governmental structures controlling the flows of inputs and outputs (e.g., through deregulation, dismantling, devolution, privatization, etc.). The question we explore below is what conditions support an escalation of the debate from a focus on internal structures to one involving restructuring of the state.

At least two consequences derive from the contentious and contradictory politics of balancing the input-output demands on the state from capital and labor. The first, noted by Habermas (1975) and Offe (1984), among others, is that state policies invariably constitute short-term, opportunistic (i.e., suboptimal) tactical responses to political demands rather than strategic solutions to underlying problems. This suboptimal nature of policy outputs has been widely noted to explain the high frequency of compromised regulatory policies, weak or nonexistent enforcement mechanisms, substitution of rhetoric for substance, and similar manifestations of the politicization of state policy.

A second and less widely noted consequence of these contradictory politics of state-labor-capital relations is that the most radical redirection of inputs and outputs—a redirection manifested as government restructuring—is most likely to occur under conditions of a congruence (real or perceived) of interests among these otherwise conflicting and contradictory parties. That is, a redirection or reconcentration of state resources toward *either* capital *or* labor is possible without risk of the corresponding input crisis only when such a policy shift can be supported by a congruence of interests of both capital *and* labor.

Now, the possibility of such congruence might appear to be nonexistent in light of the contradictory and mutually exclusive nature of demands on state resources emanating from capital and labor. But such an expectation relies on an overly passive view of state actors, in which the state inertly receives demands for outputs and compliantly responds to threats of withheld inputs. The politics of input-output conflicts does not just happen to the state. State actors work opportunistically and instrumentally to manipulate, orchestrate, deflect, and control the substance, magnitude, and timing of inputs and outputs obtained from and demanded by capital and labor. State actors, in other words, intervene to influence public perceptions and understandings of rights and responsibilities, where rights are defined as expected state outputs and responsibilities define expected inputs to the state (see, for example, the chapters by Cope and Herod in this volume). The instrumentality through which state actors exert their influence differs according to the form of the state and the system of organization of the state apparatus. In an authoritarian or totalitarian state, for instance, such influence can be exerted directly by fiat. In a liberal democracy, state actors lack such direct control over economy and society and they rely, instead, on discursive manipulation through the use of rhetoric and the harnessing of ideology (see, for instance, Bachrach & Baratz, 1970; Bowles & Gintis, 1986; Lukes, 1974).

In rejecting an overly passive view of the state as simply responding to contradictory demands (and avoiding countervailing threats) from capital and labor, it is equally essential not to overestimate the instrumental capacity of state actors to orchestrate political outcomes at will. What is required is the *convergence* of (1) an underlying structural economic crisis and (2) a political cadre (either within or outside the state apparatus) that is poised and able discursively to control the definition and meaning of the crisis for those affected by it. Extreme examples of such convergence might include the New Deal's first hundred days, Germany between the wars, or China under Mao. In all of these cases, pervasive economic crisis was discursively represented by a charismatic leader in a form that mobilized an otherwise fractured polity in support of radical state action.

It is further possible to argue that the severity and duration of the resultant redirection of state policy—the extent of state restructuring—depends on the severity and reach of the underlying economic crisis. Where the preponderant part of a national polity is substantially and negatively affected by structural economic crisis and falls under the sway of the state actors' discursive control, as in the cases just cited above, there is less potential, both numerically and politically, for groups or individuals not affected by crisis conditions to introduce an alternative reading into the public debate or to seek an alternative outcome. Less pervasive crisis conditions allow the interjection of other viewpoints, demands, and discursive representations that challenge and fragment the state actors' discursive political control.

I contend that such a partial convergence between underlying structural economic crisis and political instrumentalism characterized the period in the United States beginning with the national economic recession of 1990-91, encompassing the congressional mid-term election of 1994, and culminating with the budgetary reductions and deregulation achieved by the Republican-controlled Congress in 1994-96. During this period, the Republican Party in the U.S. House of Representatives succeeded in marshaling a number of important and usually disparate constituencies behind a rhetoric of government deregulation and dismantling. These groups—large and small scale capital, the newly disaffected white-collar middle class, and displaced, deindustrialized blue-collar workers—were all experiencing the harmful effects of fundamental structural economic crisis. Although each of these groups experienced the crisis differently, they at least temporarily accepted the Republicans' rhetorical claim that such effects were situated not in the crisis of a capitalist market economy but in the inadequacy and failures of big government to rectify the effects of that crisis. House Republicans were able to exploit this temporary convergence to strengthen their control of the state apparatus in the 1994 mid-term election—that is, they won control of the Congress.[2] In the following section, I briefly sketch out some of the conditions that made each of these groups vulnerable to the consequences of economic crisis and susceptible to the Republicans' representation of those consequences as situated in the government, rather than in the economy.

■ Economic Restructuring and Political Response

The decade of the 1980s encompassed the longest peacetime economic expansion in U.S. history.[3] The substantial aggregate economic growth achieved during this extended boom, however, obscured worsening conditions for those individuals and sectors not positioned to benefit from the structural economic changes fueling the expansion. Increases in worker productivity were not matched by an increase in wages. Median family income grew more slowly between 1979 and 1989 than during any other postwar business cycle, the real hourly wages of 80% of the workforce declined, and income inequality increased dramatically. The real incomes of the richest 1% of U.S. families grew by 62.9% between 1980 and 1989 while the incomes of the bottom 60% of families declined in real terms over the decade (Mishel & Bernstein, 1993, pp. 48-49; O'Loughlin, this volume).

These disparities accelerated during the national recession of 1990-91. The onset of recession meant that the effects of the business-cycle downturn were added to the long-term structural effects that were already strongly in evidence during the so-called expansion of the 1980s. Total civilian employment, which had been growing at a modest annual rate of nearly 2.0% between 1979 and 1989, turned negative with a –0.2% annual growth rate between 1989 and 1991 (Mishel & Bernstein, 1993, p. 226). Unemployment among blue-collar workers that persisted through the 1980s boom accelerated in the 1990-91 recession and was compounded by an above-average rise in white-collar unemployment, an indication that capital's sectoral shift from manufacturing to services no longer promised a solution to falling profits. White-collar workers whose incomes had risen dramatically during the 1980s now joined blue-collar workers on the unemployment lines. Indeed, the growth in unemployment during 1990-92 was greater in absolute terms among white-collar workers (1.13 million additional unemployed) than among blue-collar workers (1.08 million) (Mishel & Bernstein, 1993, p. 223). Real hourly wages that dropped by 10.5% during 1977-89 continued to erode, decreasing a further 2.2% during 1989-92 (Mishel & Bernstein, 1993, p. 133). Inflation-adjusted median family income decreased by 4.4% between 1989 and 1991 (Mishel & Bernstein, 1993, p. 31).

The 1990-91 recession thus constituted a period of convergence in the dismal circumstances of both capital and labor. As capital's expansion turned negative, white-collar incomes plummeted, and blue-collar wages continued to erode. The national components of this convergence are documented by O'Loughlin (this volume). Below, I sketch the underlying economic conditions encountered by both capital and labor in the early 1990s, using data from New Jersey as indicative of national trends. As argued above, these conditions produced a temporary circumstance in which the usually disparate interests of capital and labor converged in support of the Republicans' proposals for government restructuring.

Large Capital

The primary sources of crisis confronting national and international capital are competition with other capital and wage pressure from labor (Harvey, 1978). The solution sought by U.S.-based multinational corporations throughout the postwar period has been to accelerate the circulation of capital through various strategies of capital mobility. As has been widely documented, these strategies of accelerated capital mobility take at least three distinct forms (Harvey, 1982):

1. *Geographical mobility,* involving accelerated disinvestment and abandonment of unproductive regions and relocation to more profitable locations, first in the suburbs, then to the sunbelt, and most recently, abroad;
2. *Sectoral mobility,* involving massive deindustrialization and reinvestment in the financial, real estate, and service sectors;
3. *Organizational mobility,* encompassing the wave of mergers, acquisitions, and corporate concentration accompanying the transformation of capital out of fixed investment in land, factories, and machinery and into the negotiable paper of financial holding companies.

The total structural transformation of the New Jersey economy during the postwar period is emblematic of these national strategies of crisis avoidance. Between 1969 and 1992, New Jersey lost nearly 40% of its manufacturing employment (more than 350,000 jobs) through the combined effects of geographic, sectoral, and organizational restructuring.

The *geographic shift* is evidenced by the changing proportion of U.S. manufacturing jobs located in New Jersey. While New Jersey claimed

5.5% of total U.S. manufacturing jobs in 1943, this share dropped to 3.0% by 1992. The shift is most notable in the durable goods sector. During the economic boom of 1982-89, New Jersey lost 19.5% of its durable goods jobs while the United States gained 11.6% in durable goods employment (Hughes & Seneca, 1992). Whether through transfers or through New Jersey closures and startups elsewhere, manufacturing jobs left the state as capital sought more profitable locations.

The *sectoral shift* is evidenced by the state's massive deindustrialization. While 55.4% of New Jersey's jobs in 1943 were in manufacturing, by 1992 this proportion had dropped to 15.8%. During the 1982-89 economic expansion, New Jersey lost 84,200 manufacturing jobs, but gained 612,200 jobs in services (Hughes & Seneca, 1993). This transformation led seasoned observers to characterize the 1980s as "an economic leap forward of remarkable dimension" (Hughes & Seneca, 1995, p. 6) at the conclusion of which "the state's economy was virtually reinvented" (Hughes & Seneca, 1991, p. 2).

The third mobility strategy pursued by large capital has been *organizational restructuring*. This involves several forms, including mergers of relatively equal-sized firms, acquisitions of small firms by larger ones, and the divestiture of unprofitable (or less profitable) units (O'Neill, 1996). In addition to such formal strategies of reorganization, large corporations have pursued various forms of "strategic alliances" among competitors, subcontractors, and suppliers (Harrison, 1990) and the "re-engineering" of production processes and relationships (Meckstroth, 1994). The corporate benefits of such organizational restructuring include revenue growth from access to new products and markets as well as substantial cost savings achieved by eliminating redundant facilities and workers. An additional benefit is generated through the huge fees obtained by the financial and legal consultants retained to facilitate the restructuring, which account in large part for the rebound in the financial services sector following the Wall Street crash of October 1987 (Margolis, 1995). The organizational restructuring of large corporations has paralleled the sectoral shift in capital investment described above, from manufacturing in the 1980s to banking and finance (Bellanger, 1996; Spiegel & Gart, 1996), health care (Reardon, 1995), transportation, and services (O'Neill, 1996) in the 1990s. This shift was greatly facilitated by the deregulation of these corporate sectors, in another example of the state helping to provide the conditions for capital accumulation.

The strategies of geographic, sectoral, and organizational restructuring provided temporary solutions to falling corporate profits and fueled the 1980s economic expansion. But as suggested above, the 1990-91 recession was evidence that these solutions had lost their efficacy. Between February 1989 and September 1992 (the recession began earlier and lasted longer in New Jersey than in the nation as a whole), New Jersey lost an additional 131,600 manufacturing jobs, continuing and accelerating the 30-year process of deindustrialization. But contrary to the earlier period, the recession also saw a loss of 134,100 jobs in the service-producing sector, eclipsing the loss of manufacturing jobs in absolute terms and wiping out 22% of the job gains achieved during the 1982-89 expansion (Hughes & Seneca, 1993). The shift of merger and acquisition activities into the service sector accentuated the inability of services to make up for jobs lost in manufacturing, as increasing consolidation, mergers, and acquisitions in banking, finance, transportation, and health services prompted a wave of cost-cutting layoffs in those sectors. The recent merger of Chase Manhattan and Chemical Bank, for instance, produced 12,000 job cuts from the closing of redundant facilities (Spiegel & Gart, 1996).

Confronted with stark evidence that the solutions of the 1980s no longer applied, large capital pursued every avenue to maintain the flexibility and mobility conducive to seeking out the most profitable locations, sectors, and organizational forms. In this context, large capital was both instrumental in, and highly responsive to, the political agenda of government dismantling and deregulation proposed in the Republican Party's *Contract with America* and subsequently promulgated through legislative actions. In short, the Republican platform constituted a roadmap for reducing barriers to the mobility of capital. This entailed both reducing taxes on investment and eliminating regulations. Environmental regulations in particular were targeted as barriers to capital mobility. Obvious examples include eliminating restrictions on investment expansion on public lands, gutting air and water quality standards applicable to new investment, suspending regulations that make corporations liable for clean-up costs even after they abandon contaminated industrial sites, and the like (Natural Resources Defense Council, 1996). Large capital was highly attuned to the Republicans' advocacy of "enhancing ... economic liberty" and "breaking down unnecessary barriers to entry" through deregulation, decentralization, and government dismantling (Gillespie & Schellhas, 1994, pp. 126-128).

Small Capital

Small capital faces greater constraints and fewer choices than those available to large multinational corporations. Compared to large capital, small businesses—those with 100 employees or fewer—are more locally dependent (Cox & Mair, 1988). Local dependence results from personal ties, the burden of history, the cost of moving, and ties to local supplier and customer networks. Small capital is literally fixed in production and is less able to transform itself into more liquid capital that can be repositioned in more profitable locations, sectors, or organizational forms. Small capital, in short, is left behind, facing extremely difficult economic circumstances without the option of mobility, whether geographic, sectoral, or organizational.

Some evidence of these circumstances is provided in a recent survey of New Jersey small businesses.[4] The survey interviewed companies with 100 employees or fewer in five specified two-digit SIC categories (paper, chemicals, rubber and plastics, primary metals, and fabricated metals)—the heart of the state's traditional manufacturing economy. The survey data reveal the lack of mobility available to these businesses. More than half of the surveyed firms (52.4%) have operated at their present location for more than 25 years, and one-third (33.3%) were established before 1960. It is likely that these firms have remained in the region for a variety of personal and corporate reasons that prevented them from joining the exodus of manufacturing employment from the state that is documented above.

Firms that remained in the region through necessity rather than design are likely to be in difficult economic circumstances, a conclusion that is supported by data from the survey. More than half of the surveyed firms (54.8%) either lost employees or did not increase in size between 1990 and 1995. In addition, firms lost production workers at the highest rate among all occupational categories: 33.8% of firms lost production workers compared to only about one fourth of firms losing professional, technical or managerial workers (26.1%) or sales and administrative workers (24.6%).

These firms' production characteristics reveal their dependence on the regional economy. Most of the firms (77%) produce intermediate goods that are used in the production of other goods and most (83%) engage in specialized short-run production that changes frequently in response to the needs of individual customers. Almost half of the firms (47.5%) purchase half or more of their supplies in the region and a fourth of the firms (24.4%) rely on the New York-New Jersey metropolitan region for half or more of their sales.

The local dependence of small businesses is extremely problematic. The decimation of manufacturing in the state means that these firms have lost their customer base and have experienced the disruption of their supplier networks. Businesses that in many cases supported a family for several generations have become unprofitable or, at best, have had to resort to cuts in production workers to maintain a marginal profitability. There is little if any expectation of future growth.

There are few options available for small capital facing extremely difficult economic circumstances. Small business owners are unable to blame large capital for their plight. Large capital has long since moved on and is, in any case, buffered from responsibility for its effects by the pervasiveness of free-market ideology. The resulting personal frustration and anger constitute a potent political force directed against the state as the cause of economic crisis. Declining profit margins with no expectation for improvement make government requirements and regulations an intolerable burden of time and money seen as deflected from the immediate requirement of the firm's survival. Generalizing from personal experience, small business owners indict government regulations—and not the unceasing search for greater profitability—as the burden that drove capital from the region. These circumstances contribute to the convergence of interest by small and large capital in the elimination of state regulations and big government and make both attuned to the Republican rhetoric of deregulation and government dismantling.

Labor

Labor is left to flounder in the vortex created by the increasingly drastic means pursued by large and small capital to stave off the impending profitability crisis. The consequences for labor are evidenced in the brutal data of job loss, declining real wages, loss of job security, mounting unemployment and underemployment, and a deteriorating quality of life stemming from capital's desperate search for profits via, repeating Hirsch's words, the "violent restructuration . . . of every sphere of society."

These effects have caused mounting economic hardship for larger and larger segments of the population. By the early 1990s the average U.S. worker worked longer hours for lower wages and fewer benefits than 10 years previously. As documented by Mishel and Bernstein (1993, p. 171), for the United States as a whole, "the entry-level hourly wage of a young male high school graduate in 1989 was 22.4% less than that for the equivalent worker in 1979, a drop of $2.11 per hour." The decade also saw a significant increase in workers earning less than poverty-level wages. Whereas in 1979 only 4.1% of the workforce earned wages at least 25% below the poverty level, by 1989 that proportion had risen to 13.2% of the workforce (Mishel & Bernstein, 1993, p. 147). The dramatic rise in less-than-poverty-level work was directly associated with the rise in service-sector employment, since the largest amount of job growth over the decade was concentrated in the two lowest paying industries—retail trade and services (Mishel & Bernstein, 1993, p. 173; see also Nord & Sheets, 1992). Capital's attempts to reduce wage and benefit costs resulted in a rapid increase in the number of part-time, temporary, and contingent workers, so that 5 million workers were involuntary part-time workers in 1989 (Mishel & Bernstein, 1993, p. 231). The proportion of the private workforce covered by a pension plan dropped from 50% in 1979 to 42.9% in 1989 (Mishel & Bernstein, 1993, p. 155) and the percentage of the unemployed receiving unemployment insurance dropped from 76% and 67% in the recession years of 1975 and 1976, respectively, to 40% in 1990 and 1991 (Mishel & Bernstein, 1993, p. 225).

The political implications of such conditions were clearly manifest by the early 1990s, as large segments of the workforce sought to assign blame for their worsening economic plight. At that juncture, House Republicans were poised to offer their discursive interpretation that placed blame squarely on big government. This interpretation was highly instrumental for the party seeking to win control of the government, for in order to make a claim on positions of political power, one must first establish that existing governmental structures (and their occupants) are defective—hence the pervasiveness of the rhetoric against political incumbents, big government, and "the failed policies of the past." That large elements of labor would be susceptible to such rhetoric was presaged by the corps of "Reagan democrats," who supported the Republican agenda through the 1980s, and received more recent expression in Republican presidential candidate Pat Buchanan's virulent brand of anti-government populism. At the extreme, the definition of government as the source of economic woes swells the political fringe groups and militias that explicitly define the government as the enemy.

■ Conclusion

The temporary solutions that fueled the 1980s expansion turned out to be highly selective in their effects even during the boom. The onset of recession meant that they had ceased to work entirely. While large capital benefited enormously from the expedients of geographic, sectoral, and organizational restructuring during the 1980s, the recession abruptly signaled that these temporary solutions had run their course. Large capital joined small capital and labor in the experience of negative economic growth; and the plight of these latter groups, already grim even during the period of expansion, worsened even further. All of these groups—large capital, small capital, white-collar and blue-collar labor—now shared the common experience of economic decline, if in different ways and for different reasons.

As occurs in periods of crisis, each of these groups was susceptible to the representation of the problem contained within the rhetoric of the failure of government and the necessity of state restructuring as the solution. For the political actors jockeying for positions of power, the dismantling of big government was a means rather than an end. Both political parties adopted a strategy and rhetoric of government restructuring in order to mollify their respective core constituencies, who were being buffeted by massive economic dislocation, hoping thereby to secure and expand their hold over positions of political power and their control of the apparatus of government.

The form of state restructuring envisioned by Democrats and Republicans differed sharply because each party spoke to different core constituencies, whose demands on the state differed sharply as well. Republicans proposed decentralization and deregulation as ways of redirecting the benefits of government structure in favor of corporate capital. Their calls to restructure the government to unshackle the free enterprise system allowed them to retain the support of their business constituency while expanding their appeal through the promise of job creation and economic growth. Democrats called for downsizing government as a means to justify their inability to direct more government resources toward the needs of labor and communities and to avoid alienating those constituencies and losing their hold on the presidency. Both parties relied on broad ideological claims and the power of rhetoric to obscure the targeted nature of their programs and to broaden their appeal. In short, government restructuring was a means for both parties to manipulate the structure and resources of the state to reward their respective political supporters and thereby to consolidate their hold on political power.

Politics in the 1990s is largely a process in which these three intersecting elements—the structure and organization of the state, the distribution of state resources to varying constituencies grappling with economic crisis, and the political fortunes of politicians—are simultaneously debated and determined. In the absence of a more convincing alternative vision, otherwise conflicting constituencies acceded in 1994 to the Republicans' interpretation of underlying economic crisis as a crisis of state structure and supported a solution defined in terms of radical government restructuring. An alternative to that vision depends less on a change in economic circumstances and more on the persuasiveness of an alternative interpretation of the implications of economic crisis for the role, function, and organization of the state. Contrary to the incantation of the Democratic bumper sticker, it is not just "the economy, stupid," but rather the prevailing interpretation of its implications that generates popular political support for restructuring the state.

NOTES

1. A previous version of this chapter was presented at the Annual Meeting of the Association of American Geographers in Charlotte, NC, April 1996. Thanks to Lynn Staeheli, Colin Flint, and members of the Rutgers University Project on Regional and Industrial Economics for helpful comments on earlier drafts.

2. The 73 freshmen Republican Representatives elected in November 1994, whose apparent unwillingness to compromise their demands for deregulation and government dismantling has proven to be a problem for Republican House leadership, are individuals who both appear to believe the Republicans' discursive representation of economic crisis as a problem requiring state restructuring and now are in positions of political authority wherein they are able to act on their beliefs.

3. The 1980s expansion (specifically, the 92 months from November 1982 to July 1990) was exceeded in duration only by the 106-month economic boom of 1961 to 1969, which was spurred by federal spending during the Vietnam War.

4. The mail survey of New Jersey small businesses was conducted in Spring 1995 by the Center for Urban Policy Research as part of a larger project, funded by the New Jersey Department of Environmental Protection, aimed at designing tax incentives for pollution prevention. The project surveyed small companies (100 employees or fewer) required by the New Jersey Pollution Prevention Act to conduct pollution prevention planning. The 88 completed interviews represent a response rate of 20.9%. (For additional information on the survey methodology and findings, see Lake, 1996).

REFERENCES

Bachrach, P., & Baratz, M. (1970). *Power and poverty.* New York: Oxford University Press.
Barrow, C. (1993). *Critical theories of the state: Marxist, neo-Marxist, post-Marxist.* Madison: University of Wisconsin Press.
Bellanger, S. (1996). Choose your partners: The era of banking consolidation. *Bankers Magazine, 179,* 19-23.
Bowles, S., & Gintis, H. (1986). *Democracy and capitalism: Property, community, and the contradictions of modern social thought.* New York: Basic Books.
Carnoy, M. (1984). *The state and political theory.* Princeton: Princeton University Press.
Clark, G., & Dear, M. (1984). *State apparatus: Structures and language of legitimacy.* Boston: Allen and Unwin.
Cox, K., & Mair, A. (1988). Locality and community in the politics of local economic development. *Annals of the Association of American Geographers, 78,* 307-325.
Cushman, J. (1995, November 25). E.P.A. is canceling pollution testing across the nation. *New York Times,* p. 1.
Cushman, J. (1996, May 26). Congress likely to repeat many environmental disputes as new budget takes shape. *New York Times,* p. 16.
Cypher, J. (1980). Relative state autonomy and national economic planning. *Journal of Economic Issues, 14,* 327-349.
Gillespie, E., & Schellhas, B. (Eds.). (1994). *Contract with America: The bold plan by Rep. Newt Gingrich, Rep. Dick Armey, and the House Republicans to change the nation.* New York: Times Books, Random House.
Habermas, J. (1973). What does a crisis mean today? Legitimation problems in late capitalism. *Social Research, 40,* 39-46.
Habermas, J. (1975). *Legitimation crisis.* Boston: Beacon.
Harrison, B. (1990). The return of the big firms. *Social Policy, 21,* 7-19.
Harvey, D. (1978). The urban process under capitalism: A framework for analysis. *International Journal of Urban and Regional Research, 2,* 101-131.
Harvey, D. (1982). *The limits to capital.* Chicago: University of Chicago Press.
Hirsch, J. (1978). The state apparatus and social reproduction: Elements of a theory of the bourgeois state. In J. Holloway & S. Picciotto (Eds.), *State and capital: A Marxist debate* (pp. 57-107). London: Edward Arnold.
Hirsch, J. (1981). The apparatus of the state, the reproduction of capital, and urban conflicts. In M. Dear & A. Scott (Eds.), *Urbanization and urban planning in capitalist society* (pp. 593-607). New York: Methuen.
Hughes, J., & Seneca, J. (1991). One-half century of job growth in New Jersey: 1940 to 1990. *Rutgers Regional Report, Issue Paper No. 2.* New Brunswick, NJ: Edward J. Bloustein School of Planning and Public Policy, Rutgers University.
Hughes, J., & Seneca, J. (1992). The New Jersey manufacturing employment hemorrhage. *Rutgers Regional Report, Issue Paper No. 4.* New Brunswick, NJ: Edward J. Bloustein School of Planning and Public Policy, Rutgers University.
Hughes, J., & Seneca, J. (1993). Anatomy of a recession: New Jersey's long road back. *Rutgers Regional Report, Issue Paper No. 6.* New Brunswick, NJ: Edward J. Bloustein School of Planning and Public Policy, Rutgers University.
Hughes, J., & Seneca, J. (1995). New dimensions of national and regional output and productivity: New Jersey's economic history revisited. *Rutgers Regional Report,*

Issue Paper No. 12. New Brunswick, NJ: Edward J. Bloustein School of Planning and Public Policy, Rutgers University.

Lake, R. (1996). *Tax incentives for pollution prevention in New Jersey.* Report prepared for the New Jersey Department of Environmental Protection. New Brunswick, NJ: Center for Urban Policy Research, Rutgers University.

Lukes, S. (1974). *Power: A radical view.* London: Macmillan.

Margolis, D. (1995, June 30). Mergers setting pace to break record of 1994: Deals in first three quarters total $564 billion worldwide, 43% ahead of last year. *Los Angeles Times,* p. D-2.

Meckstroth, D. (1994). Reengineering U.S. manufacturing: Implications of structural changes in the U.S. economy. *Business Economics, 29,* 43-49.

Mishel, L., & Bernstein, J. (1993). *The state of working America, 1992-93.* Economic Policy Institute Series. Armonk, NY: M. E. Sharpe.

Natural Resources Defense Council. (1996). *Damage report: Environment and the 104th Congress.* New York: Author.

Nord, S., & Sheets, R. (1992). Service industries and the working poor in major metropolitan areas in the United States. In E. Mills & J. McDonald (Eds.), *Sources of metropolitan growth* (pp. 255-278). New Brunswick, NJ: Center for Urban Policy Research Press.

Offe, C. (1984). *Contradictions of the welfare state.* Cambridge, MA: MIT Press.

O'Neill, P. (1996). The trend of aggregate concentration in the United States. *American Journal of Economics and Sociology, 55,* 197-212.

Reardon, L. (1995). The restructuring of the hospital services industry. *Journal of Economic Issues, 29,* 1063-1081.

Solo, R. (1977). The need for a theory of the state. *Journal of Economic Issues, 11,* 379-386.

Solo, R. (1978). The neo-Marxist theory of the state. *Journal of Economic Issues, 12,* 829-842.

Spiegel, J., & Gart, A. (1996). What lies behind the bank merger and acquisition frenzy? *Business Economics, 31,* 47-52.

Wolfe, A. (1977). *Limits of legitimacy.* Glencoe, IL: Free Press.

2

Economic Globalization and Income Inequality in the United States

JOHN O'LOUGHLIN

The federal government, long viewed as a key actor in the amelioration of poverty, is now increasingly seen as a causal element in the trend toward increasing income inequality in the United States. As noted in other chapters in this book, the size and scope of the national government are now being questioned and changed at the same time as the U.S. economy is more permanently embedded in the world economy. Growing inequality in the United States is occurring despite, or maybe because of, strong macroeconomic factors, such as low unemployment rates and high GDP (gross domestic product) growth rates.

In this chapter, I will examine the hypothesis that the increasing globalization of the U.S. economy, through increased trade and immigration, is an important element in the development of inequality and, by implication, in the role of national government. I first present the complex and contradictory evidence related to the debates over income inequality in the United States. These debates usually focus on trend data, but the data are often for different time periods, industrial sectors, and income measures. As a result, the "hard evidence" presented by individual researchers is contradictory. These contradictions have been used by politicians and others to support their own political positions and their views on whether income inequality is increasing or decreasing.

In the second section of the chapter, I present the key arguments over the causes of economic inequality in the United States. This debate is delineated between those who emphasize processes at the domestic scale and those who identify processes of economic change at the global scale. I conclude that the evidence is strongest for the role of economic processes related to globalization, but that these processes interact with domestic changes rather than stand in isolation. Once again, I pay close attention to the political biases within the debate.

In the third section of the chapter, I make the case that, while economists treat the U.S. economy as a unit, geographers are more careful to differentiate the relative impact of globalization on the fortunes of regional and local labor markets in a rapidly changing global economic environment (Agnew, 1988). This geographic differentiation adds an important dimension to the common worry about the development of a "dual-society"; therefore, I will discuss the different impact that globalization has on various labor markets of the United States. The geographic perspective identifies the way that the creation of economic and social spaces recursively interacts with economic inequality. Geographies of opportunity and disadvantage are both a cause and a product of the economic inequality of social groups within countries.

I conclude by considering the political implications of these economic developments for the dual role of the state in promoting accumulation of national capitalists while maintaining the legitimation of the state through the implementation of measures that try to cope with the adverse effects of globalization on communities and industrial sectors. Diverse economic goals across space and differential economic opportunity across social strata encourage political factionalism across both spatial and social lines.

■ The Evidence for Growing Inequality in the United States

In this section, I provide the context for my later argument by outlining the trends in income inequality using a variety of statistical indicators. The political nature of this debate obfuscates the picture, for those wishing to promote a rosy view of the U.S. economy and its benefits to all segments of society are prone to use a different set of economic indicators than more pessimistic commentators. I begin by setting the historical context of the current debate and then present the statistical indicators used by the optimists, followed by those adopted by the pessimists. To conclude this section, I show that controlling for educational qualifications is necessary to illuminate the fact that income inequality has indeed increased since the early 1970s.

Trends in income inequality are typically examined over the past quarter-century, since about 1970. This starting date is very important, because the early 1970s mark the beginning of the global economic downturn (Wallerstein, 1979). The people of the United States became accustomed to world leadership and increasing domestic prosperity in the quarter-century after 1945. By contrast, the past 25 years have been characterized by economic anxiety, a "social and cultural war" about basic American values, and belated concern about a widening income gap between rich and poor. The mass of indicators measuring American decline give mixed signals; although most economic indicators show relative decline, the cultural and military-political indicators show continued U.S. dominance (O'Loughlin, 1993). Elliott (1996) makes a persuasive case that the nostalgia now visible in the United States for the halcyon days of the 1950s is misplaced, because these golden years will not return. Instead, he argues that the more appropriate comparison is with the 1900-1914 period, when, as now, immigration was high, the economy experienced frequent oscillations, and semi-isolation reigned in the political sphere. It seems that future expectations matter more than present status. National surveys show great pessimism about the next generation. Robert Samuelson (quoted in Pearlstein, 1996) argues that much of today's economic anxiety reflects not so much a decline in what Americans *have* as an increase in what they *expect*. In this sense, we see a frustration of rising expectations of wage and economic security.

Benjamin Disraeli's comment about "lies, damned lies and statistics" comes quickly to mind as one slogs through the morass of evidence on the scope and extent of social inequality in the United States. Different statistical indicators are adopted by commentators depending on the picture they wish to paint. The statistical debate can be easily seen in the dispute about the use of a key index, median family income, which has been stable or slightly falling in real terms over the past quarter-century (O'Loughlin, 1993). In interpreting this income figure, it should be remembered that the "median family" has changed, with both more single-parent and dual-income households ("Politics Into Economics," p. 25). Furthermore, it matters whether one considers only wage income or also includes non-income indicators. Non-income indicators show that Americans are living better than ever. Using more general measures of well-being such as amount of free time, number and quality of consumer goods, real cost of basic necessities, and health and educational measures, the population generally is better off than ever before, even at the bottom of the income spectrum (Pearlstein, 1996). But measures of wage income are mixed. It matters, for example, if one makes calculations using per capita income (optimists point out that real per capita income is up 38% since 1973) or median family income, used by pessimists and those who wish to demonstrate the existence of inequality (Pearlstein, 1996).

Bearing in mind these warnings about changes in family structure, what the data measure, the politically charged nature of the debate, lies, and statistics, we can turn attention to the central question: What is the evidence related to income trends? As noted above, the evidence is mixed and often confusing. The remainder of this section examines the statistics related to income in the United States and attempts to interpret them in light of their political implications.

Evidence of "meritocratic inequality" (Bluestone, 1994, p. 87) is compelling; the traditional U.S. emphasis on equal opportunity, not equal outcome, shines clear. There was never a majority for equal outcome, though there has been a national agreement for equal opportunity for over 20 years. Politicians across the political spectrum seem to agree with the Federal Reserve Bank in Dallas, which takes the position that "inequality is not inequity" and that "America as a Land of Opportunity lost . . . is just plain wrong" (quoted in Pearlstein, 1996, p. 6). That this position is no longer unique to the United States is seen in the recent retreat from the "equal outcome" policy in the United Kingdom to the "equal opportunity" policy of the United States. Other responses to the worries about growing inequality are to challenge the data, to argue that inequality can change quickly in a society of high social and employment mobility like the United States, and to dismiss inequality as irrelevant in a time of high economic growth (Krugman, 1994c, p. 140).

The pessimists, who make the case of an increasingly unequal society, use indicators that show that (a) the United States is the most "unequal" of the set of rich countries; (b) median family income in the United States has been stagnant for over two decades; (c) while the rich get richer, the relative gap with the poor grows; and (d) growing inequality is happening despite greater productivity per worker in all sectors, especially in the manufacturing sector. Using Gini Coefficients of income, the United States is the most unequal of all rich OECD countries ("America: Still World Capital of Inequality," 1996). Unlike other countries with high inequality scores, the rich in the United States are better off than anywhere else (based on the 90th percentile as a percentage of median income) except Ireland. However, the poor (the bottom 10th percentile) in the United States are significantly worse off than in any other country. The welfare net is lower (and will get lower with passage of the 1996 Welfare Reform Act) in the United States than in any other rich country except Portugal (measured as welfare spending as a percentage of GDP). According to this analysis, U.S. wage inequality (the ratio of the lowest wage decile to the median wage) grew by 15% between 1980 and 1995.

Median family income in the United States rose steadily from 1950 to 1975 but has been stagnant for the past 20 years, while GDP continues to grow at the same rate as the 1950s and 1960s. Fourteen percent of new jobs created in the most recent expansion are in the "help-supply" services, up from 5% in the early 1980s expansion (David, 1995). This kind of "out-sourcing" relieves big companies of fixed costs but pushes more people into inferior service jobs and adds to income inequality. U.S. Census Bureau data provide the evidence for growing income inequality by quintile. The richest 20% of the population now owns 48.5% of the income pie compared to 40.6% in 1969, whereas the poorest 20% have gone from 5.6% in 1969 to 3.6%. The middle categories are quite stable, but the percentage of the income enjoyed by the top fifth of families is the highest ratio since the 1920s. We may not know exactly what caused the dramatic developments of the past two decades in the standard of living for Americans, but we know that it began in the early mid-1970s; the 1980s (the Reagan years) was a time of relative stability in wages. The relative level of manufacturing wages remains an important issue because this sector still contributes 70% of U.S. exports (Fry, 1995).

To clarify the statistical exchange over the pattern of U.S. income inequality, levels of education and real wages within sectors must be considered. Only then can we conclude that income inequality is growing and that the 1970s was the climactic decade. Indeed, there seems little doubt about growing inequality in the United States when one compares income quintiles or examines individual circumstances while controlling for educational qualifications. Between 1963 and 1987, the ratio of earnings of college graduates to high school dropouts grew from 2.11 to 2.91. In the 1980s, the real income of high school dropouts fell by 18%, that of high school graduates fell by 13%, while that of master's (6+ years of college) graduates increased by 9%. Women fared better than men in overall wage growth. Three of four U.S. workers have not finished college, so that, in total, only 15% of the workforce has seen increased wages in the past decade. More than half of all workers with a high school diploma or less have experienced a loss of income in the past decade (Mishel & Bernstein, 1993; see also Kosters, 1994). With the trend of falling real wages (7% decline since 1973, according to Levine, 1995, p. 91), the United States is increasingly "penniless" but with low unemployment because of its "flexible" labor market (Krugman, 1994b).

One can also compare real wages for the same sectoral workers in the same sector as Leamer (1996a) has done for production workers from 1961 to the present. The evidence shows dramatically that the 1970s was the decade of great change. By examining the 90th, median (50th), and 10th percentiles (highest, average, and lowest paid workers), Leamer has identified the developments in real wages for production workers. At the 10th percentile, wages for production workers increased from about $4.50/hour in 1961 to $5/hour in 1971, remained at that level in 1981, but increased to about $6.50/hour in 1991. The median wage earner was at $6.75/hour in 1961, $7.10/hour in 1971, remained there in 1981, and showed a dramatic shift to $9.50/hour in 1991. The highest paid production workers moved from $8.50 in 1961 to $10/hour in 1971, increased to nearly $11/hour in 1981, and showed a dramatic increase to nearly $15/hour in 1991. In this sector, at least, it seems clear that there has been a dramatic increase in income inequality.

The economic predicament of the American worker is only fully captured when productivity is considered as well as income. Real wages within most sectors are stagnant at the same time that worker productivity is growing (David, 1995). In the past decade, U.S. productivity growth has skyrocketed. Radical changes have occurred in the American way of business; and vast numbers of people have been affected by "downsizing," "re-engineering," "delayering," and "creative destruction." The World Economic Forum now ranks the U.S. economy as the most competitive and U.S. workers as the most productive in the world. Profits of companies have again risen after stagnancy in the early 1980s; but these profits have been disproportionately turned over to shareholders or corporate executives or used for further investment. Compensation for workers has been relatively flat, and the gap between compensation and productivity continues to widen. The business mantra of the day is "productivity." As Krugman notes, "Productivity is not everything, but in the long run, it is almost everything" (quoted in David, 1995, p. 2).

Consideration of productivity suggests that the growing embedment of the United States in the global economy is a key component of the income inequality debate. Once again, however, this is a contentious issue, for there are some scholars who argue that globalization is to blame for growing income inequality and others who point to changes within the domestic economy as the culprit. In the next section, I will summarize the debate between these two competing arguments.

■ **Globalization and Rising Income Inequality in the United States**

Given the evidence for the growth of income disparities in the United States, various attempts to account for it have failed to come to any sort of consensus on its causes. The main cleavage is between those who believe the culprit is the changing nature of the domestic economy and its shift from manufacturing to service jobs and those who emphasize the increased incorporation of the American economy into the global markets through trade, immigration, and investment. Freeman and Katz (1994) have tried to account for the various domestic and international factors leading to rising wage inequality in the United States, and their list has been extended by Bluestone (1994). Of the ten possible "culprits," nine can be considered as "globalization" factors, derived from the growing integration of the United States into the world economy or sectoral shifts due to the changing nature of jobs as a result of employment relocations. Two components of domestic change are often cited as causes of income inequality: technological change and deunionization. In this section, I review the empirical evidence for and against each of these arguments related to the domestic economy. After showing that there is no conclusive evidence for these arguments, I will turn to the processes of globalization, namely, the trade deficit, factor price equalization, and immigration. Again, I review the empirical evidence for and against each position.

The Domestic Economy

The loudest advocate of the *changing technology* explanation for U.S. inequality is Paul Krugman. Like world-systems theorists, he believes that "in 1973, the magic went away" (Krugman, 1994c, p. 3). Krugman is honest in his assessment that the causes of rising inequality are still unclear. He argues that the new information technologies tilt the earnings distribution by rewarding skilled, highly educated labor, while reducing the demand (and therefore, the wages) for the products of the uneducated and unskilled workers (Krugman, 1994a, b, c, 1995; Krugman & Lawrence, 1994; Krugman & Venables, 1995). Even within the same sector, there is a widening gap between the top and the bottom of the educational spectrum. As machines replace workers, consumers are buying relatively fewer goods and more services (Krugman & Lawrence, 1994).

There is little empirical support for the technological change thesis presented by Krugman. There are few signs that the rate of innovation (new machines or new products) is increasing (Bluestone, 1994). Most businesses, for example, are not introducing technologies that require new skills; if anything, there has been a de-skilling of tasks. The formerly low-level secretarial job of typing has been decentralized by word processing to all employees. Furthermore, the impact of technological change on income inequality varies widely depending on the form of the statistical model and the type of data used, rather than on the technological change itself (Leamer, 1994).

Related to the technological explanation is the *deindustrialization* argument, advanced by Barry Bluestone (1994) and Borjas and Ramey (1993), among others. This explanation appears to have more empirical support. The high service ratio in the United States, now over 75%, has important inequality repercussions since the wage gap in this sector is large. In the manufacturing sector, the ratio of earnings between high school dropouts and college graduates moved from 2.11 to 2.42 between 1963 and 1987, while the ratio moved from 2.20 to 3.52 in the service sector (Bluestone, 1994; Kosters, 1994). Recent employment growth has come in the services sector, with cities like New York losing 600,000 manufacturing jobs while gaining over 700,000 service jobs between 1953 and 1984 (Castells, 1988). Further evidence for the effects of deindustrialization on the wage status of residents in the labor market can be clearly seen in Detroit, which lost 67,000 automobile manufacturing jobs (at 26% above the average wage for the United States) and gained 72,000 service jobs (at 4% above the average wage) in the two decades between 1970 and 1990 (Deskins, 1996). Deindustrialization has its bright side because it allows up-skilling of jobs in other sectors (Krugman, 1994a); manufacturing wages are now only 10% higher than those of the nonmanufacturing sector when one considers the number of hours worked per week.

The *deunionization* thesis claims that as the rate of worker unionization in the United States has plummeted since the 1960s to 13%, the ability of unions to pursue their consistent position of narrow wage differentials has been undermined (Freeman & Katz, 1994). Unions have not been very successful in penetrating either the new flexible (postfordist) manufacturing sectors or the new service economy. Earnings inequality in the services sector is higher than earnings inequality in the manufacturing sector when controlled for educational and skill levels. That waving the threat of moving offshore can undermine unionization efforts is recognized even by those who do not believe the movement offshore has any appreciable effect on U.S. wage levels (Krugman, 1995, p. 242). Bluestone (1994, p. 91) thinks that U.S. labor law is deliberately inimical to the organizational efforts of unions and needs reform; stronger unions would help to redress the negotiating balance in favor of wage earners (see Herod, this volume).

Globalization

It is increasingly common for rising inequality to be blamed on *economic globalization,* which has two related elements: the persistent U.S. trade deficit (Bluestone, 1994; Prestowitz, 1991) and "factor price equalization." Increased U.S. imports have contributed to the decline in manufacturing, the sector that helped to restrain earnings inequalities by paying higher-than-average wages. A significant portion of import surplus into the United States is composed of products made by low-skilled and modestly-skilled labor in Asia and Latin America, depressing the relative wages of U.S. workers at the bottom of the skills distribution. Krugman (1995) has tried to undermine this thesis, showing that the U.S. terms of trade have not changed in the past 25 years and that most of the increased imports are not from low-wage countries. Further, American consumers benefit from lower import prices and can then spend disposable income on other goods and services. Krugman stresses the accounting identity: domestic production = domestic consumption + exports − imports. Growing imports of manufactured goods is almost matched by growing exports in most manufacturing countries; consequently, the impact of trade on the size of the domestic manufacturing sector is small. By Krugman's (1995) calculations, the trade deficit accounts for no more than one-tenth of the decline in the number of U.S. manufacturing jobs. Lawrence and Slaughter (1993) state boldly that international factors had nothing to do with America's wage performance in the 1980s, a position also supported by Bhagwati and Dehejia (1994), who conclude that increased U.S. trade did not hurt wages in the 1980s, despite the theoretical claims of factor price equalization and the contradictory empirical data. Contrary to expectations, the relative prices of imported, unskilled-labor goods rose in the 1980s, rather than falling as the theory would predict.

Factor price equalization theory offers a theoretical explanation for the globalization hypothesis. Since the mid-1970s, world trade has expanded, despite the rise in nontariff barriers in the United States and other countries. Under the wing of GATT (General Agreement on Tariffs and Trade) and now the WTO (World Trade Organization), the global neoliberal trading regime is triumphant. Formerly autarkic countries like China and the former Soviet Union have entered the world trading system (in 1978 and 1989, respectively). According to factor price equalization theory, without intervention of the states to control imports, there will be equalization of wage rates across the globe, even in the absence of multinational capital investment or low-wage worker immigration. Factor price equalization is expected to continue as trade barriers fall, transport costs are reduced, communications improve, and the newest innovations in production techniques diffuse worldwide. As factor price equalization develops (e.g., wages in the United States will become more similar to those in China and Mexico in the same unskilled categories), earnings inequalities grow because wages in high-skill jobs are not subject to the same global downward wage pressures. Leamer (1996a) estimates that free trade will reduce the wages of unskilled U.S. workers about $1000 per year, a development spurred by the 1993 NAFTA agreement. Expecting these wage trends, it is no wonder that U.S. unions strenuously opposed NAFTA.

An examination of the relationship between industrial wages and the population size of countries making up the world economy shows vividly the pressures for factor price equalization (Leamer, 1996a). The percent of the populations in rich countries like the United States and Western Europe with wages above $9/hour (in 1985 dollars) is tiny compared to the massive percent in countries like China and India with wages in the range of $1/hour and less. Leamer (1996a, p. 1) states that "if this is a global labor pool, it is a very strange one indeed, with the liquid piled high at one end and hardly present at the other." Trade barriers are, of course, one reason why the situation persists; but with falling trade barriers under the new WTO regime, the pool is expected to develop the same depth everywhere (factor price equalization).

The support for the globalization hypothesis (considering both trade and factor price equalization) is bolstered by the strong temporal correlation between hourly manufacturing wages and trade dependence for the United States. Using both CPI (Consumer Price Index) and PPI (Producer Price Index) deflators, Leamer (1996a) compared real wage trends with the increased exposure of the U.S. economy to trade (imports + exports/GDP) since 1960. The trend lines show an abrupt halt in 1973 to the steady rise in real wages. This happened *at the same time* as the United States experienced a rapid rise in trade dependence, from 7% to 15%. Prior to 1973, the exposure of the United States to trade (about 9%) was lower than the Soviet Union in the same period; the ratio by 1980 was 21% (Morici, 1995/96). Using wage data, Leamer (1993, 1994, 1996a) demonstrated that the effects of globalization are significantly greater than technological change in statistically explaining the changes in wages between 1961 and 1991. In the 1970s, the wages of unskilled workers in the United States fell by 40%; whereas in the 1980s, they rebounded by 20% as a result of the change in U.S. producer prices (Leamer, 1996b).

Finally, *immigration* is also often cited by politicians and commentators as a key cause of depressed wages for native U.S. workers. A more accurate picture is that immigration is just another consequence of the deeper integration of the United States into the world economy and plays a relatively minor role when compared to factor price equalization. There are also, undoubtedly, strong regional and local effects in the influence of immigration on prevailing wages. Cities that act as major destinations for immigrants will experience greater wage competition between native and immigrant unskilled workers. The economic processes of income inequality are embedded within the creation of new geographies of globalization, including immigration flows (see Wright, this volume).

In the 1980s, 38% of net population growth was contributed by immigration (Morici, 1995/96). A recent estimate by the U.S. Census Bureau indicates that of the 18 million new jobs to be created in the United States in the next 20 years, 13 million will be filled by immigrants, mostly from Asia and Latin America. The effect on income inequality will be exaggerated if the majority of the immigrants are low-skilled workers and compete with the unskilled native population for the shrinking pool of decent low-skilled jobs. The average legal immigrant to the United States today has one year of education *less* than the native worker. Because undocumented immigrants probably have an even larger education gap, there seems little doubt that immigrants have the effect of increasing the supply of unskilled labor in some U.S. cities, thereby depressing wages and increasing resentment among native unskilled workers in cities such as Miami (Nijman, 1996), Los Angeles, Dallas (Hicks & Dixon, 1996), and other big cities. The U.S. situation stands in contrast to Canada, where immigrants have, on average, over a year of education *more* than native workers as a result of a national immigration policy that stresses human capital skills over family reunification, as is the case in the United States (Bluestone, 1994).

Summary and Evaluation

The cause-by-cause analysis of the processes that I have discussed gives some indication of each one's role in the trend toward growing income inequality in the United States. However, a more informative picture is provided if the interaction of the processes operating at the domestic and the global scales is considered. In trying to estimate the effects of different factors in accounting for rising inequality in the United States, Freeman and Katz (1994) compared two groups of male workers in the 1980s, those with high school and those with college education. Technological change accounted for 7% to 25% of the change in respective wages; deindustrialization for between 25% and 33%; deunionization for about 20%; trade and immigration (globalization) for 15% to 25%; and finally, the trade deficit accounted for 15% of the relative changes in wages. These estimates for the 1980s show that *every major economic trend affecting the United States at the present time contributes to the growing inequality of the society.* Economic processes at both the global and domestic scales are adversely affecting income inequality. With fewer and fewer institutional constraints on market forces and the Great Society tradition in rapid retreat in the 1990s, we can expect both inequalities and social unrest to worsen.

■ A Geography of Inequality in the United States

The discussion of growing income inequality and the economic processes causing it has, so far, assumed that the U.S. economy is a homogeneous unit of analysis. In this section, I will show that processes of globalization require a conceptualization of the U.S. economy as a mosaic of regional and local labor markets. Each of these labor markets tries to compete within a global economy through the definition of geographically specific attributes.

The changing geography of production over the course of American history led, by the time of World War II, to a recognizable economic core and periphery in the United States. Geographers and regional economists distinguished between an industrial core in the Northeast and Midwest and a less-urbanized periphery in the South and West. Political allegiances were similarly defined, and a dichotomous view of the geography of the country was sufficient for many purposes. Old regional divisions began to ebb after 1945 with the industrialization of the periphery, and especially after 1970, when the fordist industrial structure of the traditional manufacturing heartland began to collapse, most notably in the steel, automobile, and chemical industries. By the 1980s, it was recognized that local distinctiveness had replaced sectional or regional divisions as the most visible element of the industrial geography of the United States (Agnew, 1987, 1988). The internationalization of the U.S. economy had led to the collapse of the traditional integrated production sectoral firms that heretofore dominated a region. Manufacturers contracted out operations, frequently to companies in other countries; some moved offshore or to a more "competitive" location in the United States. A new polarization developed between the growth regions of the West and South and the traditional heartland of manufacturing, and it was paralleled by a more localized polarization within metropolitan areas as "citadels" developed in many downtown areas as the command centers of the new business services, banking and financial operations, and multinational manufacturing. A growing social polarization in the "dual cities" of prosperity and decline became visible in most U.S. metropolises (Mollenkopf & Castells, 1991; O'Loughlin & Friedrichs, 1996).

Geographers consider the United States in the past two decades not just as a market and production point in the world economy, but as a set of local labor markets with divergent fortunes. For purposes of illustration, contrast Detroit and Boston. Detroit has witnessed a dramatic loss of automobile and ancillary manufacturing jobs to other parts of the United States and abroad—a 46% decline between 1970 and 1990, according to Deskins (1996)—whereas Boston boomed in the 1980s as a result of high-tech manufacturing. The United States has had a longstanding comparative disadvantage in leather products, miscellaneous manufactures, apparel, primary metals, transport equipment, automobiles, and electronics while enjoying a comparative advantage in industrial machinery, chemicals, tobacco products, instruments, and fabricated metals (Leamer, 1996a). Clearly, local labor markets that have a specialization in the products with a U.S. comparative advantage will generally prosper, whereas the reverse is true for the sites of manufactures with comparative disadvantage. Labor-intensive products with price reductions change the labor demand curve to generate lower real wages for unskilled workers who live in metropolitan areas with an oversupply of unskilled workers (Detroit) while raising the wages of unskilled workers in cities with many skilled workers (Boston) (Leamer, 1996a,b).

The relative economic fortunes of metropolitan areas can be seen in the census data reported in Levine (1995). Of the 12 cities in the study, all except Boston showed a decline in the percentage of wage earners in the middle-income category ($20,000 to $40,000 in 1990 constant dollars) between 1970 and 1990, from about 40% of the workforce to about 35%. When examining the wages of workers in new jobs, 72% of earners were below $20,000 in Detroit compared to only 35.5% in Boston. In the United States, jobs in export-oriented companies pay better; and this sector is booming as the 129% gain in manufacturing exports between 1985 and 1994, far outstripped the export growth of 112% or the GDP growth of 25% (Kresl, 1995; Morici, 1995-1996). While U.S. domestic companies continue to seek cheap-wage locations by avoiding unionized cities, foreign manufacturers in the United States are more concerned with the purchasing power of the population (Grant & Hutchison, 1996).

With a regional geography of economic competitiveness added to the social inequality present in all metropolitan areas, it is no wonder that the 1993 NAFTA vote in Congress exhibited a political-geographic fault line between the North and Midwest, and the South and West (Clark, 1994). Representatives from localities likely to see job losses as a result of greater imports of products of unskilled Mexican labor strongly opposed the bill, whereas representatives from states bordering Mexico or likely exporters of services and goods to Mexico were supportive. Whereas some U.S. cities, like San Francisco, have long experienced cycles of greater and lesser involvement with the external global economy—in this case, across the Pacific Ocean (Walker, 1996)—others, like Dallas-Ft. Worth, Minneapolis-St. Paul, and Portland, Oregon, are now more integrated in the economic world beyond the American borders because the economy as a whole is less isolated and autarkic (Harvey, 1996; Hicks & Dixon, 1996; Kaplan & Schwartz, 1996; Kresl, 1995).

The challenges and consequences of globalization described here for the United States are common to rich countries. The significant decline in GDP growth rates for capitalist countries after 1970, coupled with the accelerated growth of foreign direct investment to a level in the late 1980s that was 10 times that of the early 1970s, suggests a marked break with the past (Magdoff, 1992). In a longer temporal perspective, the years around 1970 mark another important breaking point. The nature of capitalism in the first 70 years of this century, with manufacturing based on fordist principles, required a sort of codependency of workers and capitalists. All rich capitalist countries were marked by decreasing income inequality as higher wages led to a rise in labor incomes relative to capital incomes (Wilterdink, 1995). The states, including the United States after 1930, also increased welfare programs to reduce the excesses of social inequality. Since about 1970, this codependency of workers and capitalists has broken down as capital incomes have increased substantially relative to that of workers. Companies display few local or national loyalties as the internationalization of manufacturing weakens the generations-old linkage with place and people. The phrase "What's good for GM is good for the country" (U.S.) reflects this local nexus. There is no sign of a turnaround, and the trend poses a dilemma for governments trying to come to terms with the globalization that affects all countries. In the United States, the devolution of government functions reflects the increasing importance of local points of production. Government restructuring is promoting local initiative at the expense of federal involvement. Though reduced federal involvement may help the competitiveness and flexibility of localities in a global economy, the growing local responsibility for the economically disadvantaged raises concerns for the less fortunate members of the least competitive places.

In the rush to study globalization, we should continue to take the traditional geographic tack of examining the dialectical relations of the local and the global, while not neglecting the actions of the national government as it seeks to mediate between the losers and winners of greater global involvement. The national government has endeavored to engage in a balancing act between local economic interests (winners and losers in globalization) as well as continuing to balance the accumulation interests of American capitalists against the need for legitimation of the political-economic system that is increasingly seen as a failure by many workers who are suffering a decline in real wages and standard of living. One political-geographic solution to these dilemmas is to decentralize the operations of the national government to the states and localities so that each can pursue its own individual economic policies (see Flint, this volume). Otherwise, the government remains caught in the conflict between the free trade interests of states like California and the protectionist interests of states like Arkansas. Therefore, government restructuring is an essential response to the economic processes operating at the global, domestic, and local scales. In the next section, I conclude my argument by highlighting some of the important themes of the political debate and how they relate to government restructuring.

■ The Politics of Income Inequality

Contemporary debates about income inequality and the appropriate political response stem from the years of the "Reagan Revolution" in the 1980s. In the presidential campaigns in 1988, 1992, and 1996, social inequality at home was related to a variety of foreign policy and trade issues, indicating that changes within the global economy stimulated domestic political initiatives. Indeed, in the bigger global picture, these concerns are not unique to the United States; other rich Western countries have the same worries and show similar trends toward greater inequality, though only the United Kingdom has values as extreme as the United States (Bluestone, 1994; Equality: Opportunity knocks, 1996, p. 43; Wilterdink, 1995). The expression *"Zwei-Drittel-Gesellschaft"* (two-thirds society) aptly characterizes the present situation of inequality in France and Germany (O'Loughlin & Friedrichs, 1996) and Canada (Levine, 1995). Though the reach and level of welfare support is greater in Western Europe than in the United States, the same trend of growing neglect of the poor can be detected through a kind of "welfare fatigue" (Bluestone, 1994). All countries feel the same globalization pressures (Barnet & Cavenagh, 1994), and governments generally adopt the same policies of trying to shield their populations and companies from the negative aspects of globalization (legitimating role of government) while promoting their national industries that have a relative advantage (accumulating role of government).

Economic processes operating at both the global and domestic scales are producing income inequalities that are embedded within a geography of opportunity and constraint. The dual social and spatial manifestation of income inequality creates a problem for government policy, which aims to ameliorate poverty while encouraging regional economic initiatives. In the face of economic globalization and the consequent increase in the role of regions and localities, it is harder for the federal government to create national policies to promote industries. Yet, giving regions and states greater roles in economic policy, including welfare policies, risks the loss of national safety nets (see Cope, this volume). Thus, regional pressures on the federal government to devolve its role in accumulation raise questions about how its legitimating role will be maintained.

In considering economic policies to promote the competitiveness of U.S. industry in the global environment, two issues are paramount. If the problem is globalization, then the answer is "upgrading skill levels" and more education. It is easier to select the wrong targets (immigrants, welfare recipients, taxes, and regulations) than to start the long, slow process of adjusting to economic change. There are two obvious legitimization options for the state to pursue in the new realities of the global economy. The first option would erect trade barriers and other devices to protect "uncompetitive industries" and their workers. In the United States, there are still over a million workers in the garment, textile, and apparel industries. Though the number is shrinking steadily, their differential concentration in the Southern states and in some big cities, such as Los Angeles, generates strong local economic and political impacts as a result of layoffs. The U.S. economy is more protected than is often realized: the equivalent tariff percentage of the nontariff barriers on steel, car, and textile imports is equal to 24% (DeMelo & Panagariya, 1992). In the long term, the disparity between a U.S. position strongly supporting free trade in the World Trade Organization and a policy that is protectionist is hardly sustainable.

The local imperative of economic competitiveness in the global economy promotes the second option, education, in which the state invests in human capital to allow upgrading skills for all workers. The costs of this option are enormous, and it takes a long time to have a noticeable effect. Leamer (1996a) believes that the adverse reaction to the extension of NAFTA to Mexico in certain sectors in the United States was a telling commentary on the shortcomings of the U.S. educational system. The returns of education can be seen in the level of the "skill premium." Surveys show an increasing return for college education (compared to high school education and dropping out) over the past two decades. Only when this educational enterprise is engaged more seriously will the current wage inequality begin to stabilize or perhaps decrease. Because education is predominantly a local and state responsibility in the United States and is likely to see less rather than more national government involvement in the future (Shelley, this volume), the geographic implications are enormous; wider differences in state and local spending would further increase local and state disparities in economic welfare.

Current political discontent in the United States focuses on the role of the government in the production of wealth and the amelioration of poverty. In this chapter, I have shown that income inequality has both spatial and social manifestations that are recursively linked. In addition, government policies aimed at increasing economic competitiveness are likely to enhance geographic inequities. There is substantial evidence to suggest that, although the macro-economic indicators look fine, it is the stagnation and decline in the standard of living of a majority of Americans that is the source of anger and of disappointment with politicians of all stripes and that is the basis of the support for politicians such as Patrick Buchanan (Phillips, 1993). The swing of ideological cycles, now in a conservative phase, is clearly evident (see also Schlesinger, 1986). Over time, geographic differences between "winners" and "losers" might exacerbate sectional political rivalries to the point of generating regional parties of protest, as was the case in the United States earlier in this century as well as in contemporary Europe. Such a development would indeed turn American politics on its head.

REFERENCES

Agnew, J. A. (1987). *The United States in the world economy: A regional geography.* Cambridge, UK: Cambridge University Press.
Agnew, J. A. (1988). Beyond core and periphery: The myth of regional political-economic restructuring and sectionalism in contemporary American politics. *Political Geography Quarterly, 7,* 127-139.
America: Still world capital of inequality. (1996). *Left Business Observer. No. 71,* 4-5.
Barnet, R., & Cavenagh, J. (1994). *Global dreams: Imperial corporations and the world order.* New York: Simon & Schuster.
Bhagwati, J., & Dehejia, V. H. (1994). Freer trade and the wages of the unskilled—Is Marx striking again? In *Trade and wages: Leveling wages down?* (pp. 36-75). Washington, DC: American Enterprise Institute, 1994.
Bluestone, B. (1994, Winter). The inequality express. *American Prospect, 20,* 81-93.
Borjas, G. J., & Ramey, V. A. (1993). Foreign competition, market power, and wage inequality: Theory and evidence. *NBER Working Paper* (No. 4556). Cambridge, MA: National Bureau of Economic Research.
Castells, M. (1988). High technology and urban dynamics in the United States. In M. Dogan & J. Kasarda (Eds.), *The metropolis era* (Vol. 1, pp. 85-110). Newbury Park, CA: Sage.

Clark, G. L. (1994). NAFTA: Clinton's victory, organized labor's loss. *Political Geography 13*, 377-384.
David, P. (1995, September 16). American business: Back on top? A survey. *Economist*, p. 18.
DeMelo, J., & Panagariya, A. (1992, December). The new regionalism. *Finance and Development 29*, 37-40.
Deskins, D. (1996). Economic restructuring and social dislocation in Detroit. In J. O'Loughlin & J. Friedrichs (Eds.), *Social polarization in post-industrial metropolises* (pp. 259-282). Berlin: Walter de Gruyter.
Elliott, M. (1996). *The day before yesterday*. New York: Simon & Schuster.
Equality: Opportunity knocks (1996, August 10). *Economist*, pp. 42-43.
Freeman, R. B., & Katz, L. F. (1994). Rising wage inequality: The United States versus other advanced countries. In R. Freeman (Ed.), *Working under different rules* (pp. 29-62). New York: Russell Sage Foundation.
Fry, E. H. (1995). North American municipalities and their involvement in the global economy. In P. K. Kresl & G. Gappert (Eds.), *North American cities and the global economy: Challenges and opportunities* (pp. 21-44). Thousand Oaks, CA: Sage.
Grant, D. S. II, & Hutchison, R. (1996). Global smokestack chasing: A comparison of foreign and domestic manufacturing investment. *Social Problems, 43,* 21-38.
Harvey, T. (1996). Portland, Oregon: Regional city in a global economy. *Urban Geography, 17,* 95-114.
Hicks, D. A., & Dixon, S. R. (1996). Global credentials, immigration and metro-regional economic performance. *Urban Geography, 17,* 23-43.
Kaplan, D. H., & Schwartz, A. (1996). Minneapolis-St. Paul in the global economy. *Urban Geography, 17,* 44-59.
Kosters, M. H. (1994). An overview of the changing wage patterns in the labor market. In *Trade and Wages: Leveling Wages Down?* (pp. 1-35). Washington, DC: American Enterprise Institute.
Kresl, P. K. (1995). The determinants of urban competitiveness: A survey. In P. K. Kresl & G. Gappert (Eds.), *North American cities and the global economy: Challenges and opportunities* (pp. 45-68). Thousand Oaks, CA: Sage.
Krugman, P. (1994a, March). Competitiveness: A dangerous illusion. *Foreign Affairs, 73,* 28-44.
Krugman, P. (1994b, Summer). Europe jobless, America penniless. *Foreign Policy, 95,* 19-34.
Krugman, P. (1994c). *Peddling prosperity: Economic sense and nonsense in the age of diminished expectations*. New York: W.W. Norton.
Krugman, P. (1995). America in the world economy: Understanding the misunderstandings. *Japan and the World Economy, 7,* 233-247.
Krugman, P., & Lawrence, R. Z. (1994, April). Trade, jobs, and wages. *Scientific American, 270,* 44-49.
Krugman, P. R., & Venables, A. J. (1995). Globalization and the inequality of nations. *Quarterly Journal of Economics, 110,* 857-880.
Lawrence, R., & Slaughter, M. (1993). International trade and American wages in the 1980s: Giant sucking sound or small hiccup? *Brookings Papers on Economic Activity* (No. 2), pp. 161-226.

Leamer, E. E. (1993). Wage effects of the U.S.-Mexico free trade agreement. In P. M. Garber (Ed.), *The U.S.-Mexico free trade agreement* (pp. 57-125). Cambridge, MA: MIT Press.

Leamer, E. E. (1994). *Trade, wages, and revolving door ideas.* (NBER Working Paper No. 4716). Cambridge, MA: National Bureau of Economic Research.

Leamer, E. E. (1996a). A trade economist's view of U.S. wages and globalization. Unpublished paper. Los Angeles, CA: Anderson Graduate School of Management, University of California at Los Angeles.

Leamer, E. E. (1996b). *In search of Stolper-Samuelson effects on U.S. wages.* (NBER Working Paper No. 5427). Cambridge, MA: National Bureau of Economic Research.

Levine, M. V. (1995). Globalization and wage polarization in U.S. and Canadian cities: Does public policy make a difference? In P. K. Kresl & G. Gappert (Eds.), *North American cities and the global economy: Challenges and opportunities* (pp. 89-111). Thousand Oaks, CA: Sage.

Magdoff, H. (1992). Globalization: To what end? In R. Miliband & L. Panitsch (Eds.), *Socialist Register 1992: New world order.* New York: Monthly Review Press.

Mishel, L., & Bernstein, J. (1993). *The state of working America, 1992-93.* Armonk, NY: M. E. Sharpe.

Mollenkopf, J., & Castells, M. (Eds.). (1991). *Dual city: Urban restructuring in New York.* New York: Russell Sage Foundation.

Morici, P. (1995-96, Winter). Export our way to prosperity. *Foreign Policy 101,* 3-17.

Nijman, J. (1996). Ethnicity, class, and the economic internationalization of Miami. In J. O'Loughlin & J. Friedrichs (Eds.), *Social polarization in post-industrial metropolises* (pp. 282-300). Berlin: Walter de Gruyter.

O'Loughlin, J. (1993). Fact or fiction: The evidence for the thesis of U.S. relative decline. In C. H. Williams (Ed.), *The political geography of the new world order* (pp. 148-180). London: Belhaven.

O'Loughlin, J., & Friedrichs, J. (1996). Social polarization in post-industrial metropolises. In J. O'Loughlin & J. Friedrichs (Eds.), *Social polarization in post-industrial metropolises* (pp. 1-18). Berlin: Walter de Gruyter.

Pearlstein, S. (1996, May 13-19). Are we better off or not? *Washington Post National Weekly Edition,* pp. 6-7.

Phillips, K. (1993). *Boiling point: Democrats, Republicans, and the decline of middle-class prosperity.* New York: Random House.

Politics into economics won't go. (1996, May 11). *Economist,* pp. 25-26.

Prestowitz, C. (1991). *Trading places: How we allowed Japan to take the lead.* New York: Basic Books.

Schlesinger, A. M. (1986). *The cycles of American history.* Boston: Houghton Mifflin.

Walker, R. (1996). Another round of globalization in San Francisco. *Urban Geography, 17,* 60-94.

Wallerstein, I. (1979). *The capitalist world-economy.* New York: Cambridge University Press.

Wilterdink, N. (1995). Increasing income inequality and wealth concentration in the prosperous societies of the West. *Studies in Comparative International Development, 30*(3), 3-23.

3

Globalization and Social Restructuring of the American Population: Geographies of Exclusion and Vulnerability

JANET E. KODRAS

Investigations of the effects of globalization on the U.S. population emphasize its significant economic impacts, evidenced by a two-decade stagnation in median household incomes, the redistribution of earnings and wealth from low- and middle-income households to the most affluent, and deepening income disparities across local labor markets (see O'Loughlin, this volume). In the United States as elsewhere, the deleterious effects of globalization have been greatest among economically marginal peoples and places, unable to secure favorable position within the internationalizing competition:

> ... a new world order is emerging which is surprisingly stable in its expanding core areas, and which has ridden out the very real threats presented by inflation, debts, balance of payments deficits and economic nationalism, but which is "orderly" in part because of *a new capacity to write off regions, countries and communities that are marginal to the development of this geopolitical economy* ... the real costs of the crises of the past twenty years have fallen most heavily on certain countries, regions, and classes in the poorer regions of the "Third World" ... and on those classes and communities in the First and Second Worlds that have to live in the deadlands created by economic restructuring. (Agnew & Corbridge, 1995, pp. 192-193, emphasis added)

Although attention has focused on the divergence in life prospects between economically marginalized classes and communities relative to those succored and enriched within the present system, economic polarization is not the only divisive impact of global transformation. Less recognized, but nevertheless significant, the effect of globalization in splintering the life chances of different Americans has a distinct social dimension; for restructuring alters not only class relations but also relations defined by race, gender, generation, and other facets of social life. Specifically, market ideologies driving income disparities in the current period of rapid economic change, when combined with socially divisive doctrines such as racial discrimination and patriarchy, fracture economic life prospects along fine lines, creating multiple and overlapping forms of marginality.

Consequently, global transformation plays across an American landscape of great economic and social diversity. The impacts are geographically specific, taking particular form and affecting particular groups in places defined by their relations to the national and global political economy. Over time, each place develops a distinct local mix of material and ideological conditions that influences its ability to accommodate the shifting needs of the market and to thrive within a globalizing economy. This depends not only on the match of local productive specializations to changing demands of the larger economy, but also on the fit of the local social order to the changing needs of global capital. Specifically, mobile firms, seeking favorable locations for production within the intensifying international competition, take advantage of distinct geographic variations in workforce attributes, a spatial division of labor defined in part by the *place-specific social composition of the population,* given prevailing hiring practices that typecast minorities as working for less than whites, women for less than men, foreigners for less than the native-born, and so forth. This spatial division of labor is in turn informed by *place-specific histories in social relations,* such as local legacies of racism, patriarchy, and nativism that set the relations between social groups in the locale (Massey, 1984, 1994). As globalization rewrites production landscapes, based on this finely calibrated and historically derived spatial division of labor, complex geographies of vulnerability and exclusion, power and privilege, materialize (Jones, 1995; Seabrook, 1985; Sibley, 1995; Zukin, 1991).

Elaborating on the thesis sketched above, in this chapter I examine the recent social restructuring of the American population, cast within larger transformations of the global economy and played out across a diverse landscape of productive and social relations. The chapter consists of three sections. I first review trends in the changing social composition of the American population, with emphasis on the diverging economic circumstances of groups defined by race, gender, and generation. In the second section, I explain these trends as the result of economic transformations occurring within the social order of particular places that put some social groups at a disadvantage relative to others. I illustrate by example how the social order of a given place may render it superfluous to the needs of capital, which increasingly holds the power to write off marginal "deadlands." Alternatively, the nature of social relations in a place may render it indispensable to the needs of capital, at least for the moment. Understanding the recent social restructuring of the American population requires examining economic transformations within the uneven class, race, gender, and other social relations of particular place contexts, a confluence of economic change and social structure that alters the life chances of different societal groups.

In the concluding section, I discuss the political implications of social restructuring, as segue to following chapters. A variety of social interest groups, working at the local to national levels, have responded politically to the insecurities and frustrations emerging during economic transformation. The result is a period of great tumult in American politics, enriched by the diversity and passion of participants, but impaired by the divisiveness and rancor created by interest groups fighting for position in a volatile economy, flailing about in search of whoever or whatever is responsible for diminished prosperity and dimmed hopes.

■ Social Restructuring in the American Population

Throughout its history, the American population has been renowned for its diverse social complexion, but the degree and rapidity of changes currently underway is unprecedented. Social groups have been differentially affected by transformations in the national economy as the shift in economic base from manufacturing to services, the acceleration of labor-saving technological innovations, the replacement of secure full-time employment for contingent labor, as well as rising unemployment, stagnating wages, and a host of related changes have altered divisions of labor based on class, race, gender, and so forth (Goldsmith & Blakley, 1992). To cite just one example:

As the economy changed, the number of jobs requiring physical strength and repetitive manual labor shrank, to be replaced largely by sales and service jobs requiring social skills and personal qualities defined as attractive by the dominant culture. This expanded opportunities for groups such as middle-class women but penalized groups such as young black men. Deprived of some of the state support through work relief such as the Jobs Corps, when funding for those programs was cut, they were also now largely deprived of the opportunity for earned income as entry-level jobs in warehousing, transportation, and manufacturing moved offshore or into the suburbs. (Law & Wolch, 1993, p. 186)

I examine below the diverging economic circumstances of groups defined along three primary dimensions of vulnerability in an era of economic restructuring: race and ethnicity, gender and family composition, and generation.[1]

Race and Ethnicity

The disproportionate difficulties experienced by racial minorities in the past 20 years is inextricably linked with their disadvantaged position within a labor market undergoing fundamental transformation. For example, the subordinate position of African-Americans within the workforce rendered them highly vulnerable to economic restructuring as disadvantaged workers bore the brunt of abrupt, and often wrenching, labor market transitions (Jones & Kodras, 1990; Kodras & Jones, 1991). To illustrate, Table 3.1 presents a disaggregation of 1990 poverty rates by region and residence, across major racial and ethnic groups. As has traditionally been the case given the South's long agrarian history and tradition of racial discrimination, the poverty rate among blacks living in the nonmetropolitan South is 42%, an exceptionally high rate compared with the national average of 13.5% (Falk & Rankin, 1992; Hathorn, 1990). But the most extreme rate of all is among Latinos living in central cities of the Northeast. In New York City, for example, Puerto Ricans are substantially more likely than blacks to live in poverty and to receive welfare (Rosenberg, 1987). Tienda (1989) has shown that Puerto Ricans endured a dramatic decline in economic well-being during the period of economic restructuring, the result of their concentration in northeastern central cities experiencing intense economic dislocation, their over-representation in job sectors most adversely affected by such restruc-

TABLE 3.1 1990 Poverty Rates of Major Racial and Ethnic Groups by Region and Residence

	Total	White	Black	Hispanic
United States	13.5	10.7	31.9	28.1
Metropolitan	12.7	9.9	30.1	27.8
Central City	19.0	14.3	33.8	31.7
Outside Ring	8.7	7.6	22.2	22.8
Nonmetropolitan	16.3	13.5	40.8	32.0
Northeast Metropolitan	11.6	9.0	29.1	36.6
Central City	21.6	17.9	33.4	42.8
Outside Ring	5.9	5.2	18.2	17.7
Nonmetropolitan	10.3	10.2	*	*
Midwest Metropolitan	12.2	8.1	36.2	22.9
Central City	21.4	13.2	40.5	26.9
Outside Ring	6.2	5.7	13.3	16.3
Nonmetropolitan	13.2	12.6	28.7	19.5
South Metropolitan	13.9	10.2	28.8	25.7
Central City	18.0	12.6	30.2	28.0
Outside Ring	11.3	9.0	26.8	22.7
Nonmetropolitan	20.5	15.0	41.6	37.5
West Metropolitan	12.7	11.9	23.3	26.4
Central City	15.9	14.2	30.3	28.7
Outside Ring	10.5	10.4	14.5	24.3
Nonmetropolitan	14.8	14.0	*	29.1

* insufficient data
Source: U.S. Bureau of the Census. 1990. *Poverty in the United States, 1990.* Current Population Reports P-60 #175. Washington, DC: U.S. Government Printing Office. Table 8.

turing, and their vulnerable position at the bottom of ethnic hiring queues (see also Borjas & Tienda, 1985).

Almost as extreme is the incidence of poverty among African-Americans in central cities of the Midwest. Blacks bore the brunt of the region's industrial decline since the 1970s, due to their concentration in routine manufacturing operatives jobs, which suffered the highest casualties during restructuring (Darden, Hill, J. Thomas, & R. Thomas, 1989; Kletzer, 1991; C. Wilson, 1992). The worst crisis came in the five year period 1979-1984, when a devastating one half of all black men working in Lakefront durable-goods production lost their jobs. The Detroit case study presented below explains the process behind this pattern, describing how deindustrialization combined with discriminatory practices in this particular context to generate disproportionate poverty among African Americans.[2]

Poverty rates among whites are highest in the central cities of the Northeast, where the majority of poor whites are Hispanic. In the nonmetropolitan areas of the Northeast, on the other hand, poverty is almost exclusively Anglo. The lowest poverty rates in the country are found among whites residing in the suburbs of the Northeast and Midwest, where only 1 in 20 lives in poverty. Here thrives Galbraith's (1992) "culture of contentment." Yet, even this small poverty rate represents almost three million people, evidence that poverty transcends all boundaries.

Gender and Family Composition

Table 3.2 presents 1990 poverty rates by family composition and race, disaggregated by region and residence. The overwhelming message is that female-headed families experience unconscionably high poverty rates across all categories (Pearce, 1992; Zopf, 1989). The dramatic increase in the number of female-headed families is often cited as a reason for this disproportionate impoverishment. Although breakdown of the nuclear family has reached the point that the majority of U.S. children born at present will live at least part of their childhood in a single-parent home (Ellwood, 1988, pp. 45-46), focusing on changing family composition alone does not address *why* female-headed families are more likely to be poor once they are formed (Jones & Kodras, 1990; Kodras & Jones, 1991). Once again, transformations in the larger political economy, combined with discriminatory practices in the labor market, are prime causal mechanisms. Women's subordinate position in the labor force has left those providing sole support for their families at greater risk during the recent period of economic restructuring as disadvantaged workers bear the impact of labor market transitions. This disadvantage is multiplied for minority women.

At the same time, political realignments worsened destitution among female-headed families, as the public safety net previously supporting children without fathers was shredded and the widespread negligence of absent fathers to pay child support greatly aggravated financial stress in female-headed families. The ironic result is that the increased share

TABLE 3.2 1990 Poverty Rates by Family Composition, Race, and Ethnicity Disaggregated by Region and Residence

	Total		White		Black		Hispanic	
	FHF[1]	MCF[2]	FHF	MCF	FHF	MCF	FHF	MCF
United States Metropolitan	52.5	5.9	44.4	5.2	64.1	11.1	68.4	19.5
Central City	60.9	9.1	55.6	7.8	65.3	14.1	72.3	22.1
Outside Ring	41.7	4.4	35.6	4.1	61.1	5.9	61.1	16.6
Nonmetropolitan	56.8	9.9	51.0	8.6	67.6	27.8	68.4	25.4
Northeast Metropolitan	54.5	4.6	52.1	8.0	57.9	12.0	82.3	15.8
Central City	66.7	9.8	71.8	8.4	61.5	16.2	84.6	19.8
Outside Ring	35.3	2.5	31.1	2.5	47.3	4.2	*	7.4
Nonmetropolitan	49.1	4.5	51.5	4.5	*	*	*	*
Midwest Metropolitan	53.9	4.7	39.6	3.8	68.9	13.9	66.8	13.5
Central City	64.8	8.8	52.8	6.4	70.6	18.1	66.3	14.9
Outside Ring	32.8	2.9	29.5	2.9	52.4	*	*	*
Nonmetropolitan	53.7	8.6	52.4	8.3	*	*	*	*
South Metropolitan	53.7	6.6	39.3	6.0	66.1	10.2	59.1	21.2
Central City	57.5	8.1	46.0	7.2	63.7	11.5	54.6	23.5
Outside Ring	49.6	5.8	34.9	5.4	70.1	8.6	66.7	18.4
Nonmetropolitan	59.4	12.5	47.9	9.9	69.2	28.8	*	29.4
West Metropolitan	46.6	7.5	46.5	7.0	53.3	7.5	61.3	20.1
Central City	52.0	10.1	49.3	9.1	59.6	15.1	65.6	23.2
Outside Ring	41.9	5.9	44.4	5.8	40.4	2.3	58.0	17.6
Nonmetropolitan	55.9	9.0	54.9	8.6	*	*	*	24.9

[1]female-headed families, no spouse present, related children under 18 years
[2]married-couple families
*insufficient data
Source: U.S. Bureau of the Census. 1990. *Poverty in the United States, 1990.* Current Population Reports P-60 #175. Washington, DC: U.S. Government Printing Office. Table 8.

of poverty borne by women synchronized with the increased share of employment assumed by women, as divorce and family disruption propelled unprecedented numbers of women into meager jobs at low compensation, even as fathers withheld, and the government withdrew, assistance.

The highest incidence of poverty among female-headed families occurs in the central cities of the Northeast and Midwest. Gender, race, ethnicity, and poverty are intertwined—in central cities of the Northeast, the poverty rate of families headed by a Hispanic woman is an alarming 85%; while in midwestern central cities, the incidence of poverty is highest in black female-headed families, with fully 70% living below the poverty level. Since 1970, women have experienced substantial losses in manufacturing jobs within the old industrial core, especially metropolitan areas of the Northeast and Midwest (Kodras & Padavic, 1993); and such losses have been concentrated among minority women. The forfeiture of these relatively well-paid jobs, working in tandem with cuts in the region's relatively generous welfare system, has devastated many female-headed families living in central cities of the industrial core.

Poverty is also severe in families headed by women in the South and the West. Once again, minority status plays a variable role, with the highest rates in black female-headed families in the small towns and rural areas of the South and in Hispanic female-headed families outside the big cities of the West. And once again, transformation in the manufacturing sector is an important factor. As international competition accelerated during the 1970s, many U.S. manufacturing firms, seeking areas of low labor costs, moved to the Sunbelt, particularly the rural South and suburban West, where women registered disproportionate gains in low-wage manufacturing employment (Kodras & Padavic, 1993). Most recently, however, many of the firms using large female labor pools, such as textiles in the South and electronics in the West, have moved offshore or automated, resulting in rising unemployment and poverty rates. The case study of Silicon Valley presented below describes how the destabilized work and family conditions associated with an economic boom in electronics, combined with a distinct division of labor based on gender and nationality, generate disproportionate poverty among foreign-born women working in this particular context.

Given the attention focused on the plight of female-headed families, the impoverishment of nuclear families is easily overlooked; yet half of all poor children live with both parents (Ellwood, 1988, p. 46). Among married-couple families, the highest incidence of poverty is found in the nonmetropolitan South, notably the intact poor white families of Appalachia, and in central cities of the Northeast and West, particularly in Hispanic families. Latino poverty is atypical in that there is less difference between the poverty rate of married-couple families and female-headed families. The poverty of nuclear families is very sensitive to changes in local labor markets, fluctuating in tandem with unemployment rates due to the fact that almost half of married-couple families living in poverty have at least one full-time, full-year worker (Ellwood, 1988, chapter 4).

TABLE 3.3 Age Distribution of the U.S. Population Above and Below the Poverty Level, 1990

Percent Below the Poverty Level	AGE	Percent Above the Poverty Level
1.2	85+	1.0
2.0	80-84	1.3
2.1	75-79	2.2
2.7	70-74	3.2
2.4	65-69	4.2
3.2	60-64	4.3
3.0	55-59	4.4
3.0	50-54	5.0
3.1	45-49	6.0
4.1	40-44	7.9
5.3	35-39	8.5
7.5	30-34	9.0
8.0	25-29	8.3
8.2	20-24	7.0
8.1	15-19	6.4
9.7	10-14	6.6
11.7	5-9	6.8
13.7	0-4	6.9

Source: U. S. Bureau of the Census. 1990. *Poverty in the United States, 1990.* Current Population Reports P-60 #175, p. 3. Washington, DC: U.S. Government Printing Office.

Generation

A clear generational redistribution of poverty began in the early 1970s, accelerated during the 1980s, and has culminated in "the first society in history in which the poorest group in the population is children" (Moynihan, 1986). Table 3.3 shows the age structure of the U.S. population above and below the poverty level in 1990. Children are greatly overrepresented in the poverty population, comprising fully 40% of the poor but only 24% of the nonpoor. A phenomenal one quarter of all American children under 3 years old live in poverty.

Once again, these trends can be linked to transformations in the larger political economy. Declining real wages earned by their parents is the major determinant of rising poverty rates among children—a startling one half of *all* hourly workers under 25 years drew below-poverty level wages for a family of three in 1989 (Johnson, Miranda, Sherman, & Weill, 1991). Furthermore, the real income of young families with children fell by fully 25% between 1973 and 1986 (Consuming our children, 1988). Retraction of public assistance targeted to families is a secondary factor in the rising poverty rates among children (Johnson et al., 1991). Because children's poverty reflects the economic circumstances of their parents, their geographic distribution is a function of the transformations addressed above. Of particular note, the incidence of poverty among children is higher in rural than urban areas; and reflecting nonmetropolitan trends of the 1980s, the number living in poverty there increased throughout the decade despite overall growth in the U.S. economy from 1983 to 1989 (Johnson et al., 1991).

At the other end of the age spectrum, the elderly are underrepresented among the poor. The political clout of this demographic group has been instrumental in securing previous gains during the current round of budget cuts. The inflation-adjusted value of Social Security payments has more than held its own, whereas social programs targeted toward other age groups have been slashed and the real wage levels of the economically active population have declined (Phillips, 1990, p. 205). Significantly, the federal government's per capita expenditures on the aged are now ten-fold that spent on children (Consuming our children, 1988, p. 230). The escalating value of stocks, bonds, and real estate—wealth disproportionately held by the elderly—was a most important factor augmenting their position during the 1980s.

These trends should not obscure the fact that 10% of all elderly people in the United States live in poverty. Gender divisions are particularly notable among the aged poor, as three fourths of the poverty population over 65 years is female. This "genteel poverty" is especially acute in the small towns and rural areas of the South, where about one quarter of all elderly women live in poverty (Karpatkin, 1987, p. 216). Furthermore, although the elderly are underrepresented within the poverty population, they are strongly overconcentrated at income levels just above the poverty threshold. Thus, while only 10% of the poverty population is over 65 years old, they comprise almost 20% of the near-poor (with incomes 100-125% of the poverty threshold). The case study of the Pine Ridge reservation presented in the following section demonstrates that transformations in the U.S. economic and political systems have a unique effect within the particular generational traditions of the Sioux nation.

■ Geographies of Vulnerability and Exclusion

General patterns in the social restructuring of the American population have been described in terms of overall economic transformations in the previous section, but the processes generating these patterns become evident only when interpreted within the uneven social relations of particular places. In the accelerating international competition of an integrating world economy, mobile capital scans the global stage for the most favorable locations. Its very mobility highlights minute differences between places, throwing into sharp relief local advantages in labor supply, political interests, resources, and infrastructure (Harvey, 1990). Caught in this larger competition, each place presents a distinct local mix of material and ideological conditions that influences its ability to accommodate the shifting needs of the market and thrive within a globalizing economy. This depends not only on the match of local productive specializations to changing demands of the larger economy, but also on the fit of the local social order to the changing needs of global capital.

As seen from the perspective of mobile capital, the social order of a place includes both the social composition of its population and the social relations between groups in that population. Specifically, the *social composition of a local population* constitutes a potential labor force with particular qualifications and demands, given prevailing hiring practices that typecast minorities as working for less than whites, women for less than men, foreigners for less than the native-born, and so on. The search for competitive location is not simply a function of the physical composition of the local workforce, however, as evidenced by the fact that mobile capital has largely avoided the Mississippi Delta, despite its notoriously low-wage black workforce. Firms are also sensitive to the legacy of *relations among social groups in the locale,* such as local traditions of unionization and worker rights that set the relations between labor and capital in a place, or local legacies of racism, patriarchy, and nativism that set the relations between social groups in the locale. The relations among the various local groups are socially constructed in a place and are unique to it, but they do reflect traditions of discrimination and bias in the larger society that are refracted into a particular place context as local variations on an American theme.

Set within the context of particular places, economic restructuring can, in turn, alter the social order of those places, as changing structures of production and mechanisms of distribution reorganize occupational structures; shift the distribution of income and wealth; and transform class, race, gender, and other social relations. Thus, to understand the recent social restructuring of the American population, we need to examine economic transformations within the uneven class, race, gender, and other social relations of particular places, *a confluence of economic changes and social structures that alters the life chances of particular societal groups living within specific contexts.*

I will illustrate these points below with three brief case studies: Detroit, where the interplay of deindustrialization and racial discrimination devastated working-class black communities; Silicon Valley, where the intersection of economic boom and gender divisions of labor in the electronics industry provided jobs, but at below poverty-wage rates, to a largely foreign-born female labor force; and South Dakota's Pine Ridge reservation, where the convergence of economic restructuring, shrinking federal responsibilities, and the history of cultural genocide deepened the impoverishment of the Sioux nation.[3]

Motor City

Detroit presents a particularly notorious case of the detrimental effects of economic restructuring, as the world oil crisis, rising international competition, and global reorganization of the auto industry have ravaged its economy and eliminated fully 70% of its manufacturing jobs since the peak productive years of the 1960s (Luria & Russell, 1981). Restructuring exhibited a distinct racial dimension here, with African-American workers bearing the brunt of job loss, because deindustrialization took place within a long tradition of discrimination in Detroit's labor and housing markets. First, black workers were largely confined to the lowest level jobs in heavy manufacturing, the segment hardest hit during deindustrialization. Of particular importance, the social contract arbitrated between management and labor—The Big Three auto makers and the United Auto Workers (UAW)—used dual job ladders and seniority lists based on race, which locked in place a distinct racial division of labor (Hill, 1986; C. Wilson, 1992). Systematic bias in the labor market was aggravated by pervasive discrimination in the housing market as exclusionary zoning, neighborhood covenants, and real estate steering hindered black suburbanization, thus confining African-American workers to the aging, inner city plants (Hill, 1986; C. Wilson, 1992). As the auto industry contracted in the 1970s and 1980s, job losses and plant shutdowns were greatest in these inner city plants; and unemployment and poverty among the African-American population skyrocketed. In Detroit, fully 35% of the black population, and an extraordinary 50% of black children, lived in poverty by 1990. The detrimental impacts of deindustrialization are made human, given faces of varying hues and features, when examined within the specific context of social relations evolving in a particular place.

Although Detroit is but one place on the unfolding map of vulnerability and exclusion drawn by the processes of globalization, this brief case study illuminates processes underlying racial patterns in social restructuring reviewed in the first section above. African Americans living in midwestern central cities suffered one of the highest poverty rates in the country (Table 3.1) as massive deindustrialization, set within a distinct racial division of labor, ravaged black working-class communities in the old industrial core.

Silicon Valley

Over the past 25 years, the U.S. economy—and the diversity of social groups seeking a livelihood within it—has been buffeted by sharply contradictory trajectories of decline and growth played out in different places across the country. In striking contrast to the deterioration of heavy manufacturing in the Midwest, numerous metropolitan areas in the West and South experienced strong economic growth over the period. The archetype is Silicon Valley, centered in Santa Clara County south of San Francisco. Employment increased by 1000% over the past two decades, as the old economic base in agriculture and canning was transformed into the world's densest agglomeration of high-tech enterprises producing computers, laser and microwave equipment, electronic components, and advanced instruments (Saxenian, 1985).

Yet even in this setting, global competition and economic restructuring generated dramatically different life prospects for different social groups. First, it is not generally understood that the wealth amassed in this entrepreneurial paradise was supported, and indeed made possible, by an assembly-line labor force working for subsistence wages. More than 100,000 people live in poverty here, amidst the corporate headquarters, flash palaces, and luxury cars that compose the official image of Silicon Valley. Second, the gender division of labor on the assembly lines has shifted abruptly, as the electronics industry, caught in whitehot rivalry for international market share, has sought to reduce costs:

> a predominantly male production workforce, well paid and earning substantial benefits, was replaced with a cheaper, more expendable female labor supply beginning in the 1960s. By the late 1970s, one-half of all employment in the electronics industry consisted of subsistence-wage assembly-line jobs with few prospects for advancement; fully 75 percent of these jobs were held by women, including large numbers from Mexico,

Vietnam, the Philippines, and Korea. . . . In an area rich with the potential for making fortunes, the fast pace, workaholism, and unrelenting stress took a toll on individual lives and family stability. By the late 1970s, divorces outnumbered marriages in Santa Clara County, overcoming national and even California trends. The economic effects of family breakdown were disproportionately borne by women, due to restricted opportunities for self-support in the high technology firms of Silicon Valley. (Kodras, in press, based on Hossfeld, 1990; Rogers and Larsen, 1984; Stacey, 1984)

The result is a great gap in living standards between the rich and poor, further fragmented by gender, nationality, and marital status. Compared with the national average poverty rate of 13.5%, the rate for white married-couple families in Santa Clara County is virtually nil, whereas that for families headed by Latina, Asian, or black women ranges from 30 to 40%. Thus, the impacts of economic growth are not distributed equitably across social groups, as even fast-growth firms seek global competitiveness in labor costs by capitalizing on the uneven social relations that evolve in place over time.

The case study of Silicon Valley provides a second account of the processes underlying the intricate map of vulnerability altered by globalization trends. Female-headed families living in the metropolitan West exhibited substantial poverty rates during this time (Table 3.2), as a distinct gender division of labor, combined with destabilized work and family conditions associated with booming sectors of a globalizing economy, generated high levels of poverty among women working in high-technology plants, even as others made fortunes.

Pine Ridge Reservation

The Sioux reservation lands of South Dakota were "deadlands" long before globalization of the U.S. economy began to exert an effect, but the current round of restructuring has greatly worsened life chances for its peoples. Site of the massacre at Wounded Knee 100 years ago, the Pine Ridge reservation has been subjected to economic marginalization and cultural exploitation throughout its history, rendering it especially vulnerable to transformations in the larger political economy during the past two decades. The cumulative effect is that Shannon County, which contains most of the reservation, now holds the dubious title of poorest county in the United States—a remarkable 63% of its population lives below the poverty level in dismal comparison with the national average of 13.5%. The infant mortality rate is three times the national average; deaths by suicide, two times; by homicide, three times; and by alcoholism, ten times the national rates (Kilborn, 1992). Ninety-four percent of the population is Native American.

The bleak and deteriorating circumstances of life in Pine Ridge originate in the unique relationship between the federal government and reservation peoples. First, it is not generally recognized that the U.S. government is virtually the only service provider. Reservations are not subdivisions of the states, nor are they considered local governments, so the mechanisms used by other localities to generate a tax base, maintain infrastructure, and provide services are structured out of existence by federal decree (Hertzberg, 1982). As a result, federal policies addressing health, housing, jobs, education, and so forth are the only remaining alternative: "[F]or the tribes, the federal government isn't just the safety net—it's the whole circus. The tribes can't turn to the states for help, and reservation economies have little or no tax base for the tribal governments to tap" (Hertzberg, 1982, p.16). And the dismal state of reservation economies also originates in the orders of the federal government. Most reservations are stranded far from urban markets, a legacy of the historical U.S. policy of consigning Native Americans to economically marginal scraps of the western frontier. In addition, land is held in trust by the federal Bureau of Indian Affairs, which greatly constrains the ability of reservation peoples to create wealth and cultivate a strong local economy.

Given this overwhelming reliance on the federal government—a dependency, it must be remembered, that originates in actions taken by the state, not in the presumed indolence and apathy of the people—Native American reservations were extremely vulnerable to government cutbacks beginning in the late 1970s and 1980s. It was at this time that the effects of stagnation and restructuring of the American economy began to exert an effect on the federal government as declining tax revenues and rising inflation constrained its ability to meet former responsibilities. The Reagan administration targeted federal social policies in its cutbacks, arguing that the rising costs in social programs were the source, rather than the result, of national economic problems. As increasing numbers of Americans faced economic insecurity within their own households, many bought the argument that the government had squandered their hard-earned tax dollars on the lazy and undeserving, despite the fact that less than 4.0% of federal spending was devoted to such programs (Ellwood, 1988). The cuts had negative consequences for people in need of assistance throughout the country, but reservation populations were particularly hard hit, as once again, federal programs were the primary mechanism for providing education, housing, health, and so forth.

Following large cuts in federal funding in the early 1980s, unemployment, poverty, and social problems exploded on reservations across the country (Task Force, 1993). On the Pine Ridge reservation, per capita income declined by fully one quarter, to just $3,400 (Kilborn, 1992), and the poverty rate climbed by 20 percentage points, leaving two thirds of the population below the poverty line. These transformations in the larger political economy had a generational effect on the reservation, but the situation is more complicated than might first appear. As elsewhere, poverty rates among the young have surpassed those of the aged (Figure 3.1), as programs assisting the aged (e.g., Social Security, Medicare) were spared the sharp cuts made to programs benefiting children and their parents (e.g., WIC, AFDC, and other health, education, and job creation programs). In Shannon County, the poverty rate for persons under 18 years reached fully 70% in 1990, compared with a rate for persons over 65 years of 57%, a 13-point spread. But the measurement of generational differentials in the quality of life, using census definitions and categories, takes on new meaning here because of the communal nature of life on the reservation, with extended families living together and sharing resources (Kilborn, 1992). This last case study reinforces the point that understanding socially divisive effects of economic restructuring requires examining the processes within specific social relations evolving in particular place contexts. Although the vignettes presented here are but three places on the dynamic American map of vulnerability, an account of virtually any locale would illustrate the impacts of economic restructuring cast within particular social contexts.

■ Conclusion: The Political Implications of Social Restructuring

The intensifying competition of economic and social groups jockeying for position within the globalizing U.S. economy plays out in the political arena from the local to the national levels. The ability of different groups to secure a position is affected by their participation in both formal politics (who votes, whose interests are represented by the state) and extra-governmental activism (social movements, political violence). In recent years, a great variety of social interest groups, each in their own way threatened by the fluctuations and insecurities of a globalizing American economy, have increasingly vocalized their political discontent; and the ascendance of "identity politics" to a major role in national political discourse has altered the strategies and priorities of these interest coalitions, including a reconsideration of the appropriate role of government in a changing American society. Many of these individual threads of discontent have been woven into a pervasive challenge of the national government, under charges ranging from sinister domination to blustering ineffectiveness, from overregulation to neglect of responsibilities, from profligate spending to abandoning priorities. The *Contract with America* is but one expression of this trend toward targeting the state as the source of societal problems.

Why the state? The rancor and divisiveness that currently characterize American politics largely reflect the economic polarization and social fragmentation engendered by globalization, yet political leaders of these social movements, and indeed many politicians working within the state itself, lacking much leverage to change conditions in the global market, increasingly charge that the state is responsible for stagnating affluence and diminishing prospects. The ultimate outcome of these efforts is not yet known, but it is clear that a diminished role of the national government will feed back on the social well-being of the American population, differentially affecting the life prospects of particular groups living in specific place contexts, thus generating new rounds of social restructuring.

NOTES

1. In recent years, both the flow of income and the stock of assets have become increasingly polarized within American society as changes in the labor market, the state, and capital and property markets—all features of a larger restructuring in the economy—have reconfigured the distribution of resources, leaving an increasing number to struggle in destitution, the majority to founder in stagnation, and a select few to amass unprecedented fortunes. To document the detrimental impacts of economic restructuring, I present trends in poverty rates across various social groups. The most commonly used measure of deprivation is the poverty rate, defined as the proportion of a population living in a household with income below a given threshold. In 1990, a family of four with two children would be counted among the poverty population if total household income was less than $13,254. The rate is controversial, as might be expected given that any decision as to how poverty should be defined has political implications, and must therefore be used

with caution (see Kodras, in press). The poverty rate is used here because it is the only spatially and temporally comprehensive, yet demographically specific, measure of deprivation available for our purposes. Even this variable is unavailable for small social groups in the American population, so the impact of restructuring on their well-being is left undocumented. Unless otherwise noted, all data on poverty rates are drawn from the 1970, 1980, and 1990 *U.S. Census of Population.*

2. Note also that the poverty rate for blacks in the Midwest is more than three times as high in the central cities as in the suburbs. It may be no coincidence that William Julius Wilson (1987), the noted scholar who has drawn attention to a widening gap between well-to-do suburban blacks and poor inner-city blacks, works at the University of Chicago, where the evidence around him is particularly acute.

3. A more detailed account of these and other case studies appears in Kodras (in press).

REFERENCES

Agnew, J., & Corbridge, S. (1995). *Mastering space: Hegemony, territory, and international political economy.* London: Routledge & Kegan Paul.

Borjas, G., & Tienda, M. (1985). *Hispanics in the U.S. economy.* Orlando, FL: Academic Press.

Consuming our children. (1988, November 14). *Forbes,* p. 230.

Darden, H., Hill, R., Thomas, J., & Thomas, R. (1989). *Detroit: Race and uneven development.* Philadelphia: Temple University Press.

Ellwood, D. (1988). *Poor support: Poverty in the American family.* New York: Basic Books.

Falk, W. A., & Rankin, B. H. (1992). The cost of being black in the black belt. *Social Problems, 39,* 299-313.

Galbraith, J. K. (1992). *The culture of contentment.* Boston: Houghton Mifflin.

Goldsmith, W., & Blakley, E. (1992). *Separate societies: Poverty and inequality in U.S. cities.* Philadelphia: Temple University Press.

Harvey, D. (1990). Between space and time: Reflections on the geographical imagination. *Annals of the Association of American Geographers, 80,* 418-434.

Hathorn, C. (1990). Down and out in the Delta. *The Nation, 251,* 50-53.

Hertzberg, H. (1982). Reaganomics on the reservation. *New Republic, 187,* 15-18.

Hill, R. (1986). Crisis in the Motor City: The politics of economic development in Detroit. In S. Fainstein, N. Fainstein, R. Hill, D. Judd, & M. Smith (Eds.), *Restructuring the city: The political economy of urban redevelopment* (pp. 80-125). White Plains, NY: Longman.

Hossfeld, K. (1990). Their logic against them: Contradictions in sex, race, and class in Silicon Valley. In K. Ward (Ed.), *Women workers and global restructuring* (pp. 149-178). New York: ILR Press.

Johnson, C., Miranda, L., Sherman, A., & Weill, J. (1991). *Child poverty in America.* Washington, DC: Children's Defense Fund.

Jones, J. (1995). Beyond "race and culture": American underclasses in the late twentieth century. In G. Demko & M. Jackson (Eds.), *Populations at risk: Vulnerable groups at the end of the twentieth century* (pp. 1-18). Boulder: Westview.

Jones, J. P., & Kodras, J. (1990). Restructured regions and families: The feminization of poverty in the United States. *Annals of the Association of American Geographers, 80,* 163-183.

Karpatkin, R. (1987). Afterword. In E. Richards, C. Bird, & J. Altongy (Eds.), *Below the line: Living poor in America* (pp. 215-216). Mount Vernon, NY: Consumers Union.
Kilborn, P. (1992, September 20). Sad distinction for the Sioux: Homeland is No. 1 in poverty. *New York Times,* pp. A1, A32.
Kletzer, L. (1991). Job displacement, 1979-1986: How blacks fared relative to whites. *Monthly Labor Review, 114,* 17-25.
Kodras, J. (in press). The changing map of American poverty in an era of economic restructuring and political realignment. *Economic Geography.*
Kodras, J., & Jones, J. P. (1991). A contextual examination of the feminization of poverty. *Geoforum, 22,* 159-171.
Kodras, J., & Padavic, I. (1993). Economic restructuring and women's sectoral employment in the 1970s: A spatial investigation across 380 U.S. labor market areas. *Social Science Quarterly, 74,* 1-27.
Law, R., & Wolch, J. (1993). Social reproduction in the city: Restructuring in time and space. In P. Knox (Ed.), *The restless urban landscape* (pp. 165-206). Englewood Cliffs, NJ: Prentice Hall.
Luria, D., & Russell, J. (1981). *Rational reindustrialization: An economic development agenda for Detroit.* Detroit, MI: Widgetripper.
Massey, D. (1984). *Spatial divisions of labour: Social structures and the geography of production.* Basingstoke, UK: Macmillan.
Massey, D. (1994). *Space, place, and gender.* Minneapolis: University of Minnesota Press.
Moynihan, D. (1986, March 22). The family and the nation—1986. *America,* p. 57.
Pearce, D. (1992). The feminization of poverty. In J. Kourany, J. Sterba, & R. Tong (Eds.), *Feminist philosophy: Problems, theories, and applications* (pp. 207-219). Englewood Cliffs, NJ: Prentice Hall.
Phillips, K. (1990). *The politics of rich and poor: Wealth and the American electorate in the Reagan aftermath.* New York: HarperCollins.
Rogers, E., & Larsen, J. (1984). *Silicon Valley fever: Growth of high-technology culture.* New York: Basic Books.
Rosenberg, T. (1987). *Poverty in New York City: 1980-1985.* New York: Community Service Society of New York.
Saxenian, A. (1985). The genesis of Silicon Valley. In P. Hall & A. Markusen (Eds.), *Silicon landscapes* (pp. 20-34). Boston: Allen and Unwin.
Seabrook, J. (1985). *Landscapes of poverty.* Oxford, UK: Basil Blackwell.
Sibley, D. (1995). *Geographies of exclusion: Society and difference in the West.* London: Routledge & Kegan Paul.
Stacey, J. (1984). Sexism by a subtler name? Postindustrial conditions and postfeminist consciousness in Silicon Valley. *Socialist Review, 14,* 7-28.
Task Force on Persistent Rural Poverty, Rural Sociological Society. (1993). *Persistent poverty in rural America.* Boulder, CO: Westview.
Tienda, M. (1989). Puerto Ricans and the underclass debate. *Annals, American Academy of Political and Social Sciences, 501,* 105-119.
Wilson, C. (1992). Restructuring and the growth of concentrated poverty in Detroit. *Urban Affairs Quarterly, 28,* 187-205.
Wilson, W. J. (1987). *The truly disadvantaged: The inner city, the underclass, and public policy.* Chicago: University of Chicago Press.
Zopf, P. (1989). *American women in poverty.* Westport, CT: Greenwood.
Zukin, S. (1991). *Landscapes of power: From Detroit to Disney World.* Berkeley: University of California Press.

4

Citizenship and the Search for Community[1]

LYNN A. STAEHELI

> [T]wo concerns . . . [lie] at the heart of our discontent.
> One is the fear that, individually and collectively, we are
> losing control of the forces that govern our lives. The other
> is the sense that, from family to neighborhood to nation,
> the moral fabric of community is unraveling around
> us. These two fears define the anxiety of the age.
>
> (Sandel, 1996, pp. 57-58)

The 1980s and 1990s have been times of economic and social dislocation for many Americans, as the previous chapters have demonstrated. Americans have responded to this dislocation in many ways. Some have experienced increased levels of anxiety, fear, and sometimes anger. Others have tried to reposition themselves and their households in the new economy by working longer hours or by retraining. Still others seem to have acceded to changed economic circumstances and have attempted to address social issues. They have turned their attention beyond purely economic concerns as they search for community and a sense of common purpose that can unite Americans as citizens. Through their efforts, citizenship and community are back on the political agenda. In this chapter, I focus on these efforts as they shape the form of state restructuring. I argue that the anxiety created by economic and social restructuring has created a climate in which citizens return to long-standing debates over the nature of the Republic and the role of citizenship and community in self-government.

It seems that everyone has something to say about citizenship and community. The Heritage Foundation has given its journal *Policy Review* the subtitle "The Journal of American Citizenship." The Communitarian Movement sponsors a journal called *The Responsive Community*. A plethora of new books written by academics for general readership have been published. And politicians of all stripes have contributed their thinking on the topics. These contributions include two books by Daniel Kemmis, the mayor of Missoula, Montana; the Republican *Contract with America;* President Clinton's 1996 State of the Union Address—which mentioned community 22 times, or once every three minutes!—and a new communitarian movement that is gaining followers.

The discourse of citizenship and community is not easily characterized by labels such as "liberal," "conservative," "right," or "left." This ambiguity is part of the appeal of community and citizenship as ideals, but it is also the source of unease that many feel when they hear these terms (Garber, 1995). Sandel (1996) argues that community and citizenship speak to competing visions of democracy that have been present since the founding of the United States. Rather than a competition between right and left, the current discussion revives the debate between liberal and Republican political theory and the accompanying conceptions of rights and common good.

Contemporary political debate about citizenship and community addresses the rights of individuals, the development of civic resources for self-government, and the definition of an empowered citizenry. These concepts are also central to the current debates over restructuring the federal government. The argument for devolution, for example, is that taking government closer to "the people" allows for greater self-government. Many people believe that privatization allows for the development of civic resources. Dismantling the apparatus of government and declaring certain functions as beyond the reach of the state is held to allow for the growth of an empowered citizenry that simultaneously takes responsibility for its fate as a community and allows greater individual freedom.

Given the importance of citizenship and community to current debates, it is important to examine carefully the discourse surrounding these terms. In this chapter, I review several recent writings about citizenship and community that are directed to the general public. The review includes writings from politicians, discussions of results from opinion polls, and the offerings of "public intellectuals" writing in nonacademic outlets. I am not attempting either a comprehensive review or a random sample of this writing. Rather, I have selected materials from the popular press that represent the public discourse that surrounds the debates over state restructuring. Reliance on written materials, of course, misses some important forums for public discourse—notably television and talk radio. As Ehrenreich (1995) notes, television is where the real Zeitgeist in American society is found. Although I cannot replay sound bites and talk radio, I supplement the discussion of the written material with results of telephone surveys and interviews with community activists that I conducted in five mid-sized cities.[2] These results suggest the ways in which the public discourse is used and interpreted by Americans. I do not intend to imply that there is uniformity or agreement across all the writings I discuss or among the survey respondents, nor do I assess their arguments as "right" or "wrong." Instead, what emerges from the literature and interviews are four interrelated themes. All four themes rarely appear in any piece or interview, so what follows is my construction of the broad discourse of citizenship and community, rather than a reprise of any given article. The themes I have identified include, (a) a reaction against individualism and a rights-based culture; (b) increased attention to morality, virtue, and responsibility; (c) an ideology that community and localities are generative of citizenship; and (d) a concern about the implications of a revival of community for those who are not citizens. Taken together, the themes contribute, first, to a sense that something is wrong in American politics and society, and second, to an emerging political sensibility rooted in community that is part of the restructuring of American life.

■ Reaction Against Individualism and the Culture of Rights

Many contemporary writers express unease with an American culture that seems increasingly characterized by individualism. Labels such as "the me generation" from the 1980s express a sense that Americans have become self-centered in ways that make it difficult to understand what links the country together. Evidence that Americans are withdrawing from public life and are lacking a sense of public purpose seems to be everywhere. Seminars are offered on "self-actualization," gated communities limit access to residential areas, enrollment in private schools is increasing, and poll results show that Americans no longer identify themselves in terms of durable relations of family. Feminists, communitarians, neoconservatives, environmentalists, Republicans, and Democrats identify—and to varying extents struggle against—a society in which the "unencumbered self" has replaced individuals who are situated in relationships that shape their outlook and sense of purpose. The loss of associational life is decried in "The Strange Disappearance of Civic America" (Putnam, 1996), *Rights Talk* is exposed by Glendon (1991), and the *Snarling Citizen* is identified by Ehrenreich (1995). As a result, democracy is on trial according to Elshtain (1995), and we need to "reinvent American society" (Etzioni, 1993) and seek *The Good City and the Good Life* (Kemmis, 1995). In these writings and others, two dimensions of individualism and rights are of particular concern.

First, some writers (but not all) identify a "radical individualism" that hides the ways we are connected to each other in our society. According to these authors, it is not the case that we no longer share a common purpose or a common life, but that a focus on the individual and the ways that we structure our lives in urbanized environments hides the bonds that link individuals in families, communities, and the nation (e.g., Etzioni, 1993; Kemmis, 1990, 1995). They argue that when we focus on difference and individuality, it is easy to lose sight of the things that we do share, and obligation is diminished. For example, Elshtain argues:

> [T]he celebration of a version of radical autonomy casts suspicion on any and all ties of reciprocal obligation and mutual interdependence. What counts in this scheme of things is the individual and his or her choices. If choice is made absolute, it follows that important and troubling questions that arise as one evaluates the distinction between individual right and social obligation are blanked out of existence. One simply gives everything, or nearly so, over to the individualist pole in advance. (1995, p. 12)

Kemmis (1990) argues that most people do not really want to live in a world in which no one feels obligation to them and in which they are not obligated to others. Further, Sandel (1996) argues that most people feel disempowered by such individualism and autonomy. In an era where the economic forces that shape people's lives are increasingly removed from their control, it requires a sense of common purpose to effect positive changes in the economy. Individualism in this context is experienced as a loss of agency, and many Americans feel this loss.

Second, concomitant with the ascendance of individual over the communal has been an emphasis on rights and the procedures to protect them as the hallmark of liberty. According to Kemmis (1990), American political history can be characterized by a tension between an emphasis on rights for individuals and a concern for the moral conditions in which responsible citizens can be nurtured. During this century, the concern for rights seems to have dominated, with less concern for the development or maintenance of civic values (Etzioni, 1993). Rights, in a sense, have come to replace the moral obligations we feel to one another (Elshtain, 1995; Gans, 1995; Kemmis, 1990; Rorty, 1996).

In reaction to this situation, some authors argue that rights have been extended into areas not intended by the framers of the Constitution. One strand of this discussion argues that the rights protected under the Bill of Rights were defensive; they granted protection from governmental interference, rather than being positive statements of what an individual could do. By this way of thinking, there is a significant difference in the phrases "Congress shall pass no law" as compared to "Citizens are free to." Further, the focus on rights has deflected attention from the "rightness" of acts. Galston (1991) argues that some acts may be protected as rights (e.g., hate speech is protected as free speech), but that couching discussions in terms of rights diverts attention from the reprehensible nature of some acts and is therefore "morally incomplete."

A second group of authors wishes to put a "moratorium on the granting of new rights" (Etzioni, 1993) until it is possible to identify either the moral basis of rights or the constitutional basis of rights (Brownback, 1996). Among these authors, the concern is that an extension of entitlements has accompanied the extension of rights. Where in the Constitution, they ask, is the right to welfare? Where is the right to have a child you cannot support? Where is it written that others must subsidize immoral behavior? Many of these commentators have been involved in the current efforts at state restructuring. They seek to limit rights to those mentioned in the Constitution and to interpret them in terms of freedom from governmental interference. Representative Sam Brownback (1996), for example, has proposed a "New Contract with America." In it, he calls for a constitutional caucus that would review every federal law and program to identify its constitutional basis; if that basis could not be identified, the law or program would be eliminated.

Coming out of many of these writings is an argument that individuals must be seen as encumbered, as bound in relations of mutual obligation, and that rights must be exercised in context. This idea is new to the way many Americans think about rights. Kemmis (1990), for example, argues that Madison and other founders of the nation wanted to expand the borders of the United States to keep citizens apart so that conflict would not emerge. Now that the frontier is no longer able to keep Americans separated, Kemmis argues we must learn to exercise our rights in the context of community.

■ Morality, Virtue, and Responsibility

The ascendance of rights and a procedural notion of liberty in this century entailed a bracketing of conceptions of public good in political discourse. Yet notions of public good and virtue were also of concern to the founders of American democracy. Many commentators today are concerned that as a society, we have lost our ability to talk about morality and values, and that this inability has created a vacuum in civic life. Ehrenreich (1995), for example, argues that virtue is now relegated to the back of cereal boxes and discussions of low fiber diets; it no longer seems relevant to one's soul. Some argue that the 1960s radicalism rebelled against an oppressive and discriminatory morality but was unable to reconstruct a new morality to replace the old (Elshtain, 1995; Etzioni, 1993). Others argue that the increasingly diverse society in which we live makes morality and values contentious issues for public policy (Auster, 1994; Fukuyama, 1994; Noonan, 1994). Both sets of commentators argue, however, that a void is created when virtue and responsibility are no longer seen as necessary to citizenship and as a complement to rights.

Some commentators and organizations that seek to reintroduce morality to public discourse are promoting a morality rooted in Christianity; the Heritage Foundation, Focus on the Family, and the Christian Coalition are obvious examples of advocates of a Christian morality. In other cases, religious communities are seen as vehicles by which morality may be reintroduced, but the goal is not a specifically Christian morality. For example, Henry Cisneros (1996) advocates a greater role for communities of faith in shaping and implementing public policy, but he is careful to include examples involving a wide range of faiths. Still other commentators couch their discussion of morality in secular terms—never mentioning church or religion. Etzioni (1993), for example, argues that there is a wide range of moral values that everyone can agree on and that have nothing to do with any religious tradition. Among these values, he lists honesty, the democratic process, and a commitment to respect each other. Now some might dispute this list, but Etzioni's point is that it is possible to engage in a discussion of morality that does not rely in the first instance on religious texts or spirituality.

Most commentators agree that discussions of morality and virtue should be initiated in civil society, rather than in the state. As such, much emphasis is placed on reinvigorating the institutions in which discussions of moral issues can occur. These institutions may include the family, neighborhoods, churches, ethnic associations, public spaces—indeed, any setting that provides opportunities for individuals to gather with people different from themselves. Elshtain (1995) argues that it is in civil society that our connections with others become obvious: "Civil society is a realm that is neither individualist nor collectivist. It partakes of both the 'I' and the 'we'" (p. 9). When individuals retreat from public into privatized places, such as private schools and gated communities, they lose that connected sense of I and we.

It is not just the commentators who are expressing the need for a reinvigorated civil society. Polling data and focus groups conducted by several sources (e.g., Kettering Foundation, Knight-Ridder Corporation, and Greenberg Research) indicate that the general public no longer believes that government is accountable for its actions. What is remarkable, however, is that many people believe that the solution will be found in non-state institutions. Increasingly, we hear individuals saying that the public has to shoulder much of the blame for the state of democracy, because citizens did not become involved in the political process and allowed the government to take control (Kettering Foundation, 1991). Much of the activism around citizenship issues involves wresting control back from the government and vesting responsibility in the citizenry (Staeheli, 1994). Doing so, however, involves the development of greater civic virtue and capacity (Kemmis, 1990) and the recognition of the bonds that commit us to each other as citizens. As we are told, "It takes a village to raise a child" (Clinton, 1996). Bellah and his colleagues (Bellah, Madsen, Sullivan, Swidler, & Tipton, 1985) speak of this in terms of "the second language of commitment" that binds citizens together.

Part of this language of commitment is expressed as responsibility. While the discussion of morality, virtue, and obligation often sounds high-minded, the contemporary discussion of responsibility can sometimes sound less lofty. There is a heightened sense in the United States that some individuals are not acting responsibly and that the rest of society has to pay for the irresponsible behavior of some. Redistributive programs are seen as rewarding people who do not work, and affirmative action is sometimes characterized as rewarding people who cannot or will not make it on their own (Greenberg, 1995). In the context of economic changes that threaten the stability of middle-class households (Lake, this volume; O'Loughlin, this volume), the middle class has become angry at being forced to compensate for irresponsibility. In the surveys we conducted, respondents were convinced that the rewards for responsible behavior are diminishing (Staeheli, 1996). In turn, we see support for measures to penalize "irresponsible" behavior (e.g., cutting benefits to families that divorce, not increasing benefits when children are born out of wedlock). What is significant about some of these constructions of responsibility, however, is that they are *personal*. The *Contract with America*, for example, proposes the "Personal Responsibility Act" (Republican National Committee, 1994). The emphasis in this bill is on "personal" rather than "social" responsibility (Cope, this volume).

Other discussions, however, seemingly locate responsibility within the community. The Communitarian Movement, for example, proposes "The Responsive Community Platform: Rights and Responsibilities" (Etzioni, 1993). While the discussion is laced with references to responsibility, there is virtually no discussion of state programs. Rather, responsibility is fostered through moral discourse and teaching in the community. Many not-for-profit service organizations also argue for the benefits of clients assuming some degree of responsibility. For example, one feminist organization in Washington State makes recipients of assistance demonstrate how they will "give back" to the community. In an interview, the director of the foundation argued that this requirement is important to the self-esteem of recipients, that it helps to generate more social capital in the community, and that it is a fundamental part of being in a community.[3] Virtue and responsibility, then, are clearly linked in the goals of this organization and are held as ideals for both the community and individuals. This is a linkage that is increasingly made in the discourse of citizenship and community.

■ The Revived Importance of Community

The appeal of "community" may be the way it links individuals to the collective through the concept of "unity." In this spirit, a community organization in Colorado distributes bumper stickers proclaiming "Celebrate CommUNITY." But more than a slogan, community seems to play three roles in the new discussions of citizenship.

First, community, like civil society, is used to contextualize the exercise of rights. Communitarians of all stripes argue that absolute rights simply cannot be exercised in a social context; the exercise of rights by one individual inherently influences, and often infringes on, the exercise of rights by others. Debates over speech rights (e.g., Fish, 1993), pornography (e.g., MacKinnon, 1994), and reproduction (McDonaugh, 1994) highlight this point. Once again, the idea of the frontier in the American political culture plays an important role. Kemmis (1990) argues that Madison's vision of citizens separated in space meant that a sense of responsibility to, or awareness of, other citizens played little role in his conception of citizenship. Kemmis continues to point out, however, that the frontier is closed and that we do not have the option of moving away from our neighbors. Furthermore, he argues that in contemporary society, community, rather than separation, will provide the conditions whereby Americans can exercise their right to self-government and can shape their future as citizens.

Second, and as implied above, community is seen as generative of citizenship. It is the institution in which responsibility is fostered and rights can be guaranteed through the sense of mutual interdependence and reciprocal obligation. Some argue that it is the setting in which discussions of morality and virtue should occur. In a time when polling data suggest that the public is repelled by individualism and distrusts government and business in equal measure, community seems to be the only institution of civil society that can provide the bonds of obligation and trust that are needed for democracy to survive. A recent Knight-Ridder survey found that most Americans believe community is vital to democracy; they are seeking ways to make community viable again. Indeed, over half of the people surveyed wished to re-establish the sense of community they believe characterized the 1950s (Thomma, 1996). In my own surveys, the sense of community was not as high, but about 20% of the respondents volunteered "strengthening communities and neighborhoods" as important changes for their cities. There was a clear sense that communities had to be involved in addressing the difficulties that cities and citizens faced (Staeheli, 1996).

Third, the invocation of community as fostering democratic values, institutions, and practices is usually place-based, rather than based on communities stemming from affiliation. Communities of affiliation smack too strongly of interest groups and influence peddling. Rather, many are now arguing that it is place-based communities that will provide the basis for revitalizing democracy and citizenship. Daniel

Citizenship and the Search for Community 69

Kemmis (1990, 1995) provides excellent examples of this argument. Invoking ideas of turf and territoriality, he argues that a politics of place is possible and progressive. No matter what issue is being debated, place can be deployed as a tool to make citizens understand that they share a common interest—the place. Although Kemmis recognizes that individuals may have different views of what constitutes the public good, he argues that place can be used to remind people that deadlock threatens the very places they intend to save. The common ground that is sought in political debate is, in this sense, physical. When participants understand that they have a common stake in a place, then they are more likely to seek solutions to their problems. He writes:

> The *polis* is, first of all, the place which a certain group of people recognize that they inhabit in common. Any individual or any group within that place may wish that others did not live there, but they recognize that removing them would, in one way or another, exact too high a price. Given that fact, politics emerges as the set of practices which enables these people to dwell together in this place. Not the set of procedures, not the set of laws or rules or regulations, but the set of practices which enables a common inhabiting of a common place. . . . In this way places breed cooperation. (1990, p. 122)

Voluntarism is increasingly mentioned as a way in which the practices that breed cooperation can be built. This sort of activism is hailed as the hallmark of the early republic (Etzioni, 1993) and is also a means through which we can rebuild the practices that allow "common inhabiting of a common place." Again, the emphasis on volunteering in the local community is important. Commentators such as Kemmis, Etzioni, and Galston argue that when politics are separated from places, as they are at the federal level, there is no imperative for practices that enable common inhabiting, and the result is laws and regulations, rather than socio-political practices of cooperation. Kemmis acknowledges the difficulty posed by a globalized economy in this regard (a common stake is not shared by agents that wish to use the land), but argues that we need to rebuild place-based political practices in order to address our problems.

This argument, of course, is similar to the argument for state devolution and the dismantling of governmental programs. Communities are held to know the needs of the citizenry better than any government bureaucrat or politician in the federal government, and relatively powerless people have the greatest access to decision makers at the local level. Etzioni, therefore, argues that programs should be devolved to the lowest level possible. Let the people in their neighborhoods and civic institutions decide what is best for them. Further, he suggests that these decisions should be made with little interference from the state:

> *[F]ree individuals require a community,* which backs them up against encroachment by the state and sustains morality by the gentle prodding of kin, friends, neighbors, and other community members, rather than building governmental controls or fear of authorities. (Etzioni, 1993, p. 15, emphasis in original)

The role of the state in Etzioni's communitarian vision is to use the "expressive power" of laws to promote community-defined standards of morality. The role of the national government is to ensure that majoritarianism does not rule and that the civil rights of individuals or social groups are not infringed. Although it is unlikely that most citizens have actually read the communitarian platform, this is a sentiment often expressed by politicians in the media and citizens in focus groups and surveys (Greenberg, 1995; Kettering Foundation, 1991; Staeheli, 1996). In this way, it is believed that self-government can be achieved while still maintaining the rights of all citizens.

The current political discourse, then, locates citizenship within civil society at the local level. The focus on local communities contrasts with previous attempts to build community at higher scales, such as the nation-state (Sandel, 1996). To empower a revitalized citizenry in localities, it is assumed that decision-making and political control must be shifted from the state to civil society, generally, and from the federal government to individuals and local communities specifically. Thus, the revived attention to community and citizenship has been an important component in the way the "problem" of state restructuring has been framed and in the shaping of specific proposals for reform (see Kodras, this volume).

■ Community, Citizenship, and Membership

The emphasis on virtue, community, and citizenship in the contemporary discourse invokes ideas of people working together in a common cause and for a public purpose. It is hard to argue against such a project.

Although one might decry the moral*izing* of some commentators (Morone, 1996), moral discourse—in the sense of a discourse concerned with ethics and social justice—is usually seen as a good. Even long-time advocates of rights, such as Rorty (1996), have engaged the current discourse of community.

But the nagging question in the search for revitalized notions of community and citizenship has to do with membership. Community offers unity for those within it, for those who are members or citizens. But what about for those who are not members? For those outside the community? At some level, this would seem to be a moot issue in the United States. After all, the United States is one of a handful of Western democracies that grants citizenship on the basis of parents' citizenship *and* birth on U.S. soil. Furthermore, naturalization is an option for virtually every immigrant. In this context, membership as a legal standing seems to be available to most people living in the country. In addition, the theoretical literature on multicultural citizenship reminds us that the boundaries of the community can be porous and flexible. The book jacket for Will Kymlicka's book *Multicultural Citizenship* (1995), for instance, features Hicks' painting *The Peaceable Kingdom,* complete with animals, children, Europeans, and Native Americans. And Etzioni (1993) argues that any community that is exclusionary or that vilifies members of other communities is imperfect indeed. At a more grounded level, many community groups are working toward inclusive notions of community and citizenship. For example, in Colorado Springs, the home of the group that organized an initiative to limit the civil rights of gays and lesbians, the Citizen's Project tries to persuade people that everyone is part of the community. Their goal is to unite people in community using the idea of a "human community" that happens to be place-based. Although these attempts are laudable and represent the positive side of community, the current focus on citizenship through community has an edge to it. Shklar (1991) notes that the history of American citizenship is a struggle between inclusion and exclusion over the admissibility of members of various social groups. The current discussions of community and citizenship come at a time when many people feel under threat and when the lines between "us" and "them" are increasingly strong. It is also a time when, according to my surveys, "personal situation and concerns" is one of the most frequently cited reason for becoming involved in "community politics" (Staeheli, 1996). Here, responsibility and place-based community are important to consider in terms of membership.

First, the discussions of citizenship described above assign an important role to the fulfillment of responsibility as a condition of membership; the exercise of rights is to be tempered by the recognition of reciprocal obligation. In this context, it becomes reasonable to differentiate between those who fulfill their responsibilities to the community and those who do not. The shrill debate over welfare makes it clear that, in the minds of some, people who are poor are almost by definition failing to meet their obligations as citizens. Women who have children out of wedlock are behaving irresponsibly, as are men who fail to pay child support. It is the responsibility of parents to provide for their children; it is not the responsibility of society, and more specifically, of citizens who pay taxes. According to this logic, the failure to fulfill responsibilities as citizens, as members of the polity, allows the curtailment of citizenship rights. In an admittedly extreme case, a bill was considered in the Colorado legislature to deny voting rights in state and local elections to homeless people because they do not pay taxes. When shelters began voter registration drives, it was taken as proof that homeless people want the rights of citizenship without the responsibility (Legislation at a glance, 1996). This measure did not pass, but other proposals linking responsibility and rights will pass. For example, as Cope (this volume) notes, most of the welfare reform proposals being considered at the state and national levels include workfare provisions; that is, people will not be eligible for benefits unless they attempt to fulfill their responsibilities first through work. Gans (1995) argues that such proposals imply a narrow view of obligation and responsibility in which the community is not responsible for the welfare of its members, and in which low-income parents are forced to sacrifice their obligations to their children in favor of meeting obligation to the community. As such, he argues these proposals are part of a "war against the poor" in which the social distance between rich and poor is reinscribed through political means couched in the new language of citizenship and community.

Second, the reliance on community suggests new ways to exclude people from citizenship. Whereas segregation limits the ability of people to gain access to opportunity by virtue of residence, connecting citizenship and place carries profound implications. We see this most directly in the proposals to allow state and local governments to set residential eligibility criteria for certain programs and in the proposals to restrict the rights and access to public services for immigrants (Wright, this volume). In the context of segregated cities with varying

Citizenship and the Search for Community

abilities to meet the needs of residents and the withdrawal of high income families to "privatopias" where private provision is the rule, it would seem increasingly difficult for people who are poor or who are unable to meet community standards to gain access to particular communities that might meet their needs. And as Wolpert (this volume) notes, it is precisely those communities that have the greatest need for charitable giving that will be limited by the residents' lack of charitable capacity. Etzioni (1993) argues that all communities must share their resources, but one wonders how this might actually be accomplished.

Finally, the difficulty of reconciling values should not be ignored. It is one thing to say that a first step is to identify the common values that we do share (e.g., Etzioni, 1993), but it is quite another to find a way to build a moral community that recognizes the values that we do not share (Ehrenreich, 1995). Doing so requires a willingness to listen and to compromise that is rarely seen in any political arena these days (Kemmis, 1990). In a sense, this requires that Americans confront the diversity (with a smile, according to Ehrenreich [1995]!) that undermined the sense of mutuality and common purpose said to characterize previous eras. And although there are examples of attempts to create such a willingness to listen and understand each other, the examples of sustained success in these matters are rare (Morone, 1996). A likely outcome of emphasizing local community, therefore, may be to create fragmentation and to highlight the differences between places. Depending on the form of fragmentation and difference, the transformation of American governance may have the unintended effect of making it even more difficult to find the bonds that link citizens across places.

■ Conclusions

My reading of current discussions of citizenship and community, then, suggests an attempt to refocus political life in the United States. My reconstruction of this discourse is clearly partial—not every piece of material has been included, and this recounting undoubtedly understates the conflict and disagreements authors express on these issues. Nevertheless, a large number of people believe that Americans have lost their way and their sense of common purpose. In response, many commentators argue that the nation must search for the bonds that knit Americans together. There is a sense that we need to redirect attention to include community, morality, and responsibility, not just individual rights. The belief that this can best be done in place-based communities is widely held. Although Sandel (1996) argues that the specter of a loss of liberty represents the anxiety of the age, it seems to me that reconstructing citizenship and community is no less likely to provoke anxiety.

NOTES

1. This research was funded through National Science Foundation grant SBR-9412302; that support is gratefully acknowledged. Comments from Colin Flint and David Hodge on a previous draft of this paper are also appreciated.
2. The surveys were conducted in Pueblo, Greeley, and Colorado Springs, Colorado, and in Tacoma and Spokane, Washington. They were random telephone surveys addressing quality of life, economic change, and community activism. The surveys and interviews were conducted in 1992 and 1995. The number of people responding to the surveys and interviews was 1,949.
3. Personal communication with a community activist promised confidentiality.

REFERENCES

Auster, L. (1994). The forbidden topic. In N. Mills (Ed.), *Arguing immigration: Are new immigrants a wealth of diversity . . . or a crushing burden?* (pp. 169-175). New York: Touchstone.

Bellah, R., Madsen, R., Sullivan, M., Swidler, A., & Tipton, S. (1985). *Habits of the heart.* Berkeley: University of California Press.

Brownback, S. (1996, March/April). A new Contract with America. *Policy Review: Journal of American Citizenship,* 16-20.

Cisneros, H. (1996). *Higher ground: Faith communities and community building.* Washington, DC: U.S. Department of Housing and Urban Development.

Clinton, H. (1996). *It takes a village and other lessons children teach us.* New York: Simon & Schuster.

Ehrenreich, B. (1995). *The snarling citizen.* New York: Harper Perennial.

Elshtain, J. (1995). *Democracy on trial.* New York: Basic Books.

Etzioni, A. (1993). *The spirit of community: The reinvention of American society.* New York: Touchstone.

Fish, S. (1993). *There's no such thing as free speech and it's a good thing, too.* New York: Oxford University Press.

Fukuyama, F. (1994). Immigrants and family values. In N. Mills (Ed.), *Arguing immigration: Are new immigrants a wealth of diversity . . . or a crushing burden?* (pp. 151-168). New York: Touchstone.

Galston, W. (1991). Rights do not equal rightness. *The Responsive Community, 1,* 7-8.

Gans, H. (1995). *The war against the poor.* New York: Basic Books.

Garber, J. (1995). Defining feminist community. In J. Garber & R. Turner (Eds.), *Gender and urban research* (pp. 24-43). Thousand Oaks, CA: Sage.

Glendon, M. (1991). *Rights talk: The impoverishment of political discourse.* New York: Free Press.
Greenberg, S. (1995). *Middle class dreams: The politics and power of the new American majority.* New York: Times Books.
Kemmis, D. (1990). *Community and the politics of place.* Norman: University of Oklahoma Press.
Kemmis, D. (1995). *The good city and the good life.* New York: Houghton Mifflin.
Kettering Foundation. (1991). *Citizens and politics: The view from Main Street America.* Bethesda, MD: Author.
Kymlicka, W. (1995). *Multicultural citizenship.* New York: Oxford University Press.
Legislation at a glance. (1996, February 29). *Boulder Daily Camera,* p. 3A.
MacKinnon, C. (1994). *Only words.* Cambridge, MA: Harvard University Press.
McDonaugh, E. (1994). Abortion rights alchemy and the U.S. Supreme Court: What's wrong and how to fix it. *Social Politics, 1,* 130-156.
Morone, J. (1996). The corrosive politics of virtue. *American Prospect, 26,* 30-39.
Noonan, P. (1994). Why the world comes here. In N. Mills (Ed.), *Arguing immigration: Are new immigrants a wealth of diversity . . . or a crushing burden?* (pp. 176-180). New York: Touchstone.
Putnam, R. (1996). The strange disappearance of civic America. *American Prospect, 24,* 34-48.
Republican National Committee. (1994). *Contract with America.* New York: Times Books.
Rorty, R. (1996, June). What's wrong with "rights." *Harper's, 202,* 15-18.
Sandel, M. (1996, March). America's search for a new public philosophy. *Atlantic Monthly,* pp. 57-74.
Shklar, J. (1991). *American citizenship: The quest for inclusion.* Cambridge, MA: Harvard University Press.
Staeheli, L. (1994). Restructuring citizenship in Pueblo, Colorado. *Environment and Planning A, 26,* 849-871.
Staeheli, L. (1996). Responsibility *of* the community and responsibility *to* the community: Two discourses of morality. Draft manuscript available from the author.
Thomma, S. (1996, February 6). The '50s: Those were the days. *Boulder Daily Camera,* A1.

Part II

Transforming the Political Opportunity Structure

5

Restructuring the State: Devolution, Privatization, and the Geographic Redistribution of Power and Capacity in Governance

JANET E. KODRAS

As a pivot between Part I, addressing the complex *sources* of change in the American state and Part III, investigating the diverse *implications* of change in the state, this chapter focuses on the *nature* of state change—the strategies for restructuring government that geographically reconstitute the scale and scope of the state, redistributing power and the capacity for governance. All such strategies are inherently spatial, for changing the scale of government (via devolution of responsibilities from the national to subnational levels) or the scope of government (via privatization of responsibilities from the public sector to the nonprofit and for-profit sectors) redraws the American map of regulatory structures. The two chapters that follow in Part II elaborate on the *nature of changes in the nonprofit sector and civil society,* which are complexly woven into state restructuring.

The search for appropriate divisions of responsibility within and between the public and private sectors is an ongoing process in the American federal system, a political struggle rooted in the U.S. Constitution. Recent trends have favored reducing the role played by the national government as it sheds selective responsibilities to state and local administrations, private firms, nonprofit organizations, civic groups, and households. Determining who benefits and who loses from the current wave of devolution and privatization is speculative until after the fact, but one repercussion is already apparent: The current reductions in the scale and scope of the American state will increase geographic disparities in the role and effect of government in accordance with the capacity of institutions and groups in particular places to take on functions previously coordinated by the national government. The capacity to assume these responsibilities is defined by the extent to which fiscal resources, expertise, infrastructure, and political will exist, or can be developed, within specific locales. In this sense, capacity for governance is geographically and historically produced, reflecting the place-specific and time-accumulated material resources and discursive practices that sustain and justify actions of the state and civil society.

I develop these themes in the remainder of the chapter. First I review the primary strategies for restructuring government: devolution, privatization, and dismantling. I then place these strategies within the larger context of American federalism, characterized as an ongoing experiment in the territorial dispensation of power and responsibility. In this section, I emphasize the issues of geographic scale and scope that are inherent in any discussion of federalism but are of particular importance here, given the decibel level and potential ramifications of the debate currently raging over the appropriate role of the national state relative to other entities. The antagonists each seek to frame this debate in their own terms, using discursive practices that alternatively "decenter" or "recenter" the national state, a battle of words auguring material consequences.

I turn, finally, to a discussion of these repercussions for different places, contending that the geographic consequences of current strategies to restructure the state will accentuate the different capacities of particular locales to shoulder responsibilities shed by the national government. I conclude by arguing for the *flexible use of scale and scope,* whereby institutions and groups operating in one place draw on the resources and abilities of institutions and groups elsewhere to enhance their own capacity to meet societal responsibilities.

■ Strategies of State Change

Governmental restructuring is manifested in complex and highly differentiated ways, focusing on different facets of economic and social life (see Part III of this volume). Across all such realms, state change is currently promulgated through three primary strategies: devolution, privatization, and dismantling. Each of these is defined below as a foundation for discussion to follow.

Devolution refers to the transfer, or decentralization, of government functions from higher to lower levels of the federal hierarchy. As a transformation internal to the state that alters the scale of activities, devolution redefines government responsibilities for regulating civil society, transfers authority across levels and administrative units of government, redraws the map of government costs and benefits, and changes accessibility and entitlement to government services. In shifting responsibilities and resources to lower tiers in the federal hierarchy, the national government still retains authority to set the direction for change, as "this complex subnational reconstitution of state power and regulatory structures is occurring within a set of political, discursive, and institutional parameters established by (or mediated by) the nation-state" (Peck, 1996, p. 3). Devolution is an inherently spatial process of state change—first, because the American federal structure is a hierarchical organization of territorially demarcated governments, and second, because the uneven development of different local states generates dissimilar initiatives in response to devolution according to local needs, perceptions, and abilities.

Privatization refers to the transfer of government functions to commercial firms and nonprofit organizations, thus substituting the private sector for components of the public sector. Examples of the government's encouragement of, and entanglements with, the private sector include the establishment of quasi-governmental corporations (e.g., the U.S. Postal Service, Amtrak); the employment of private contractors (e.g., for construction of the interstate highway system and other forms of infrastructure, contracts to assess and clean environmentally damaged properties); and the use of vouchers and subsidies to be spent in the commercial sector (e.g., food stamps, agricultural export incentives, Section 8 rent subsidies).

As these examples suggest, privatization incorporates a vast assortment of institutions into the business of government. The *for-profit sector* participates through direct corporate service providers (e.g., Hospital Corporation of America, Marriott Lifecare Retirement Communities, Corrections Corporation of America); corporate-established foundations (e.g., the Ford Foundation, J. Paul Getty Trust, Lilly Endowment, Inc.); and other corporate-supported organizations (e.g., Urban League, American Enterprise Institute, La Raza, Brookings Institution). The *nonprofit sector* includes nationally known entities (e.g., the Red Cross, United Way), as well as local voluntary organizations numbering in the tens of thousands. The importance of the nonprofit sector has reached the point that it is referred to as the *shadow state,* "a para-state apparatus comprised of multiple voluntary sector organizations, administered outside of traditional democratic politics and charged with major collective service responsibilities previously shouldered by the public sector, yet remaining within the purview of state control" (Wolch, 1990, p. xvi).

Privatization redefines the scope of government but does not eliminate its role altogether, because the state externalizes only selected functions to nongovernmental entities. At issue here is the proper relationship between the public and private spheres of American life— the extent to which the state should cast off service provision while retaining sovereign authority for policy making and oversight. In common with devolution, privatization is an intrinsically spatial process of state change, particularly given the extremely fragmented composition of the nonprofit sector. As the national government privatizes many previous responsibilities and withdraws funding, local voluntary organizations become increasingly dependent on the community in which they are located, creating a spatial mismatch between communities with extraordinary needs for assistance and communities with the resources to address those needs.

Finally, *dismantling* refers to the withdrawal of a government function no longer deemed appropriate for the state to provide. Dismantling is accomplished through the outright elimination of programs or by the more covert mechanisms of cutting financial support, allowing funding to fall behind the cost-of-living, or complicating regulatory procedures to the point that administration and oversight are rendered impossible. In the case of complete dismantling, the state reduces the scope of its activities; and these either cease to exist or fall to whomever will take responsibility. Private firms may see financial incentives to acquire government assets, such as mineral extraction firms purchasing Western public lands. Alternatively, responsibilities may fall to the domestic sphere. For example, the retraction of government responsibility for long-term health care, via Medicaid, would require many households to assume responsibility for family members who are elderly or severely disabled, regardless of their financial ability or competence to do so. Because the functions discarded by the government often fall to individuals, dismantling represents the most extreme form of spatial fragmentation under conditions of state change.

These three strategies of state change reflect important differences in the understood role and acceptability of government involvement in a given function. Devolution implies that a particular policy remains an accepted function of government, but that a different tier should hold responsibility for its provision; whereas privatization indicates that a given function should be performed within society but not by government. Dismantling occurs when the function is no longer deemed appropriate, whether for program-specific or broadly ideological reasons. Such "understanding" of the legitimate role of government is time- and place-specific; competing political groups seek to impose their particular vision of the acceptable functions of government in pursuit of larger agendas to decenter or recenter the national state.

Among those who seek to check the power of the national state, *devolutionists* argue that government should be "close to the people," because the local state has greater flexibility to efficiently address local needs and effectively satisfy local preferences (Bennett, 1990; Schwab, 1988). Others seek to diminish the role of the national state through privatization: *corporatists* contend that substituting the monopoly power held by the state with market competition introduced by private firms is more cost effective and responsive to consumer demand (Le Grand & Robinson, 1984); whereas *voluntarists* assert that nonprofit organizations are most deeply rooted in communities, fostering individuals' altruistic participation in civic life (Pines, 1982; Salamon, Musselwhite, & de Vita, 1986).

Proponents of a strong national state challenge each of these arguments. The leading response to the devolutionist perspective charges that local state control creates sharp inequities in government provision because those areas with the greatest need for services have the least resources, and often the least political will, to address those needs (Smith, 1986). Even when local governments do respond to local preferences, policies enacted to serve the interests of the local majority may fail to protect the economically and politically disenfranchised. Furthermore, while devolution of authority to state and local governments may appear to support local autonomy, in practice such responsibilities can create severe problems in policy making and financing at the local level. Delegating "knotty problems" to lower tiers displaces conflict away from the national government, which may engender local political battles and fiscal crises (Cockburn, 1977).

Among those opposing privatization of state functions, critics of the corporatist perspective hold that the competition assumed to make private firms more efficient than the government is often absent, so that the state has traditionally had to take on those very societal responsibilities that private suppliers shun (U.S. House of Representatives, 1991). These critics also hold that the government must retain control over policy making and oversight of private service providers to prevent fraud and abuse by those not directly accountable to the public (GAO, 1991).

The major argument against the voluntarist position is that nonprofit organizations, highly fragmented and oriented toward amenities for elites more than services for the needy, cannot be expected to substitute for the state's entitlement provision, have no prescribed obligation to national priorities, and cannot be held culpable for their actions (Wolpert, 1993).

Finally, several participants in the debate present a more refined perspective, arguing that the determination of how best to sort responsibilities between the public and private spheres, and within the federal hierarchy, cannot be resolved in the abstract, and that the choice must necessarily depend on the particular type of policy involved. DiIulio and Kettl (1995, citing Osborne, 1993) identify four policy domains where the national state should retain control, resisting devolution to states and localities:

> *Interstate Issues*: Problems cannot be solved without federal action because state and local governments lack leverage or incentives.
>
> *Uniform Standards*: Solutions require uniform standards (e.g., social security).
>
> *Destructive Competition*: Policies are so sensitive to competition between states or localities that such competition creates negative consequences that exceed the benefits of decentralization.
>
> *Fiscal Redistribution*: Solutions require redistribution of national resources to poor regions that lack them.

Peterson (1995) elaborates on these points, contending that policies encouraging economic development are best handled at state and local levels, while redistributive policies championing the social welfare are best controlled at the national level. He finds that interjurisdictional rivalry lies at the core of this division of responsibilities: Competition between state and local governments drives them to efficiently and effectively provide basic government services and promote their economic development, but this same competition engenders a "race to the bottom" among state and local governments when held responsible for social welfare and other redistributive policies. Finally, Wolpert (1993) contributes to the debate over sorting responsibilities between the public and private sectors, finding that nonprofit organizations are well suited to contribute to amenities, such as the arts, in accordance with local preferences; but that they lack the resources and infrastructure to address critical local needs, such as social welfare, where the redistributive and risk-pooling capacity of the national government is needed.

These three primary strategies for restructuring the state, and the principles sustaining them, have been employed throughout American federal history, with each holding sway at particular points throughout this distinctive and dynamic experiment in the territorial dispensation of power and responsibility. I discuss these strategies of state restructuring in the following section that briefly traces development of the American federal system to the present moment of controversy.

■ Defining the Scale and Scope of the Federal State

As "the most geographically expressive form of all political systems," federalism is a distinctive spatial structure of governance that confers ultimate authority to no single level or branch of government as it seeks to balance power among the various components (Robinson, 1961). The American federal system is internationally distinguished for the explicit and distinct role played by governments in its subnational tiers; the total $3 trillion government budget in the United States consists of about half state and local expenditures and half federal spending (DiIulio & Kettl, 1995). In addition, almost 50% of all federal discretionary spending takes the form of grants to state and local governments (11%) or contracts to private sector organizations (37%) (DiIulio & Kettl, 1995).

Although the national government is currently the stronger partner in the American federal system, the 50 states and more than 80,000 local governments retain substantial control; and the distribution of benefits and burdens conferred by government is still largely a function of the jurisdiction in which one lives. The amount of taxes one pays, the penalty for committing a crime, the control over land use, the level of public school expenditures, the availability of public assistance, the extent of public subsidies for economic development and job growth, and myriad other regulations and rules shaping the lives of Americans vary substantially from place to place across the United States.

Alteration of the federal system, as currently under discussion, involves a geographic rearrangement of these benefits and burdens and thus a redistribution of power and resources among places. Even avowedly aspatial policies directed by the national government vary in their impact and effectiveness when implemented in different local jurisdictions (Kodras, 1990). Devolving federal responsibilities to state and local governments or shifting functions to a highly fragmented private sector greatly accentuates geographic variations in government provision.

The question of how best to divide responsibilities within the American federal system, first raised in the country's founding documents, has no final answer: "It cannot be settled by one generation, because it is a question of growth, and every new successive stage of our political and economic development gives it a new aspect, makes it a new question" (President Woodrow Wilson, cited in Malbin, 1996, p. 5). Accordingly, the power of the federal government has waxed and waned over the course of the 20th century. Its role expanded in the wake of the Great Depression in the 1930s, escalated during World War II, and increased further during the period of postwar economic growth in the 1960s. Efforts to reduce the power of the national government relative to the states grew during the Nixon administration in the 1970s, accelerated during the Reagan administration of the 1980s, and has taken on new force since the 1994 congressional election.

Over time, the national government has alternated the expansion and the retraction of its role by redefining its administrative partnerships with state and local governments (via changes in the types and magnitude of grants) and the private and nonprofit sectors (via alterations in contract activity). For example, the procurement of war supplies from private vendors and the promotion of technological innovation during World War II set the pattern of government contractual involvement with the private sector ever since. Its initial partnerships with defense and aerospace industries have expanded in the postwar era to include virtually every sector of government activity.

An important means by which the national government alters its relationship to state and local governments is changing the flow of funding, shifting between categorical grants, block grants, and general revenue sharing. *Categorical grants* (consisting of operating, capital, and entitlement programs) provide funds to state and local governments for a particular purpose designated by federal law. Specific program regulations constrain, but do not eliminate, state and local flexibility as categorical grants are implemented and administered in diverse place contexts (Pressman & Wildavsky, 1973). *Block grants* (federal stipends intended to address a general area of government activity) give subnational governments greater spending flexibility, and general revenue sharing (a lump sum payment with virtually no strings attached) concedes almost total discretion to governments at the subnational level.

Recent efforts to strengthen the power of states and localities relative to the national government have focused on shifting from categorical to block grants, although the approaches have differed in important ways (Nathan, 1996). Whereas Nixon advocated converting operating and capital grants into block grants, he retained entitlement programs, arguing nationwide responsibility to preserve a basic safety net for the American population. Reagan's position was more complicated, but ultimately followed the pattern set by Nixon. In contrast, current efforts target the conversion of entitlement grants, which provide assistance wherever and whenever need is demonstrated, into general block grants with strict limits on funding and participation. This latest proposal represents a fundamental shift in the customary role of the federal government dating back to the New Deal of the 1930s.

In this and other ways, the year 1994 signaled an important moment in the history of American federalism as the election of a Republican majority in both houses, including a large number of activist conservatives, took the lead in challenging the power of the national state. The defining document of the 1990s effort to restructure government was House Republicans' *Contract with America*. This and related efforts advocate swift reduction in the size and scope of the national government, devolving federal responsibilities to states and localities, shifting functions to the private and nonprofit sectors, dismantling regulations, and, in some cases, eliminating programs and funding altogether.

The current effort to reduce the national state is not limited to congressional initiative, however (Malbin, 1996; Pagano & Bowman, 1995). In contrast to his party's traditional defense of the national government, President Clinton seeks a "middle path" that allows decentralization and state flexibility, yet retains safeguards guided at the national level. In addition, the U.S. Supreme Court has reasserted the principle of dual sovereignty, specified in the Bill of Rights, as a foundation for several recent rulings. The 50 states have taken contradictory positions, some celebrating the prospect of increased flexibility and others fearing the loss of revenues to undertake initiatives newly expected of them.

The current trend in restructuring federalism through devolution and privatization of functions previously held by the national state—a trend identifiable in various guises for more than 20 years—has distinctive implications for governance in different places across the country. I discuss the geographic consequences of current efforts at state restructuring in the following section.

■ Local Capacity and the Geographic Redistribution of Power

Evidence from previous efforts to reduce the scale and scope of the American state suggest that the current phase will increase geographic inequities in the role and effect of government, and that the resultant patterns will correspond with the capacity of institutions and groups in particular places to take on functions formerly coordinated by the national government:

> A decade of devolution and decentralization of federal programs has already yielded considerable disparity between places because states and municipalities vary in their fiscal resources and willingness to compensate for federal cutbacks. Much of this effect could have been readily anticipated from public finance studies of fiscal federalism and the organization of the nonprofit sector. The decentralization strategy may potentially help to restrict central government growth, but it also reduces the prospects for consistently equitable remedies across America's communities by either public or nonprofit sectors. (Wolpert, 1993, pp. 37-38)

The capacity to take on responsibilities is defined by the extent to which fiscal resources, expertise, infrastructure, and political will exist, or can be cultivated, within particular localities. Over time, places take on distinctive economic characteristics (based on local resource endowments, production systems, class relations, etc.); political practices (grounded in factional ideologies, traditions of party dominance, local civic and philanthropic traditions, etc.); and social relations (defined by race, ethnicity, gender, generation, etc.). Taken together, these conditions create specific institutional contexts—accretions of organizational experience and political seasoning, capital and infrastructure—that shape the capacity of governmental and nongovernmental organizations and groups in each jurisdiction, and indeed, the local sense of what is possible and appropriate for these entities to undertake. As a result, state and local governments, private corporations, nonprofit organizations, community groups, and individual households situated in different places vary considerably in their ability to respond to state restructuring at the national level. As these differences are thrown into sharp relief by devolution and privatization, the more affluent, experienced, and innovative hold a distinct advantage (Greenberg, Popper, & West, 1991).

Although the conditions defining institutional capacity are specific to place, they are by no means generated solely *within* a given place, but are instead developed through the evolving relationship of the place to the wider world. As opposed to the traditional definition of *place* as "a bounded and static portion of space," recent conceptualizations treat place as open and mutable, shaped by forces telescoping inward as global processes are transcribed and translated into local contexts. Defining place in terms of its relations to other places, the outside as constitutive of the inside (Massey, 1994; Mouffe, 1995), suggests that institutional capacities *in* place are created as global capitalist production, national government practices, and societal conventions are differentially refracted into specific local contexts. Furthermore, institutional capacities evolve over time, "by layer upon layer of interconnections with the world beyond" (Massey, 1994, p. 8). As a result, places differ in their ability to address local needs because they are dependent on their particular position within the world economy; and although this position can change, it is heavily constrained by previous layers of investment and political experience that define the capacity to address those needs.

Seen in this light, places are highly vulnerable to alterations in the larger political economy. The current drive toward global economic integration in the midst of a two-decade slowdown in productive expansion compels change in the operation of firms and governments placed around the world as each seeks to reposition itself within the intensifying competition. The result is a fundamental transformation in the global political economy, a rearrangement of the *geographic* relations between capital, the state, and civil society.

Of the three, *capital* has become the most mobile. Although some segments of capital, such as local real estate development (see Cox & Mair, 1988), remain more place-bound and locally dependent than others, such as international finance (see Warf, 1996), capital has increased its ability to play across the global stage, situating its practices in particular places as it finds profitable sites for production (Harvey, 1982; Peck, 1995). In contrast to the growing international operation of important segments of capital, states remain largely confined to jurisdictional boundaries. *National states* increasingly operate within a global economy that they are unable to control (Clark & Dear, 1984) as nation-based fordist production, regulated by the Keynesian welfare state, yields to globally extensive postfordist production, emancipated through deregulation by the neoliberal state (Jessop, 1993; Peck & Tickell, 1994). The neoliberal restructuring of the national state, taking the particular forms of devolution and privatization in the United States, passes greater responsibility onto *local states,* which are even less capable of exerting power within a globalized economy. Finally, despite the prospect of migration for some individuals, *civil society* remains the most place-bound and locally dependent, due to material, familial, and emotional ties to community (Beynon & Hudson, 1993; Cox & Mair, 1988). David Harvey (1989) nails the point: "Labor power has to go home every night" (p. 19). These interconnected trends—globalization of capital, devolution and privatization of the state, and localization of civil society—shift power increasingly toward capital (Offe, 1985; Peck, 1995; Storper & Walker, 1989), an asymmetrical power relation termed "glocalization" by Swyngedouw (1992).

Even this cursory review of the shifting geographic relations between capital, the state, and civil society helps us to understand the current restructuring of the U.S. government as a process whereby the state renegotiates its dual role with capital and civil society. Specifically, the present restructuring of the national government in an era of economic globalization has important implications for locally based states, private nonprofit organizations, and social movements.

First, local states are left with less control to position themselves within a volatile global economy, even as the national state passes off additional responsibility for them to attempt this (Peck & Tickell, 1994). Devolution of the national state does open new "regulatory spaces" for local initiative, and the geographic consequences of this particular mode of restructuring will reflect the relative capacities of local states to respond, as discussed above. But these regulatory spaces will be constrained in all places by competitive pressures imposed by global capital and reinforced by neoliberalism in the national state (Peck, 1995). The accelerating global mobility of capital accentuates fine distinctions between places (Harvey, 1989) and exerts "economic discipline" on locales to conform to its demands (Harvey, 1985). The ensuing rivalry for investment among local states will require reducing the regulatory role previously played by government, while the taxes that underwrite governmental assistance will be kept to a minimum to ensure the area's ability to attract and retain firms. John Agnew (1987) asks the telling question: "Who dares to provide public services when the tax base may move?" (p. 188). The likely scenario, then, is a "geographically uneven drift" toward lower state involvement in the well-being of local populations (Peck, 1995, p. 224). Thus, local states subject to the demands of global capital increasingly restructure their own role to ensure that pliant labor forces and adequate infrastructural bases are available to capture and retain mobile capital. Only in those rare place-contexts where capital is overwhelmingly place-bound and locally dependent does the local state have relative autonomy within the larger economic transformations (Cox, 1991; Miller, 1994).

Second, the nonprofit sector is an insufficient substitute for the local state in this situation, given the localism and limited capacity of most voluntary organizations (Wolch, 1990; see also Wolpert, this volume). Just as the national state shifts responsibility without power to local states, it calls on the nonprofit sector to assume a greater role in social provision as it reduces financial support to those organizations (Karger & Stoesz, 1994). As the national government privatizes many previous responsibilities and withdraws funding, local voluntary organizations become increasingly dependent on the community in which they are located, and consequently, the less affluent, experienced, and innovative communities, where needs for assistance are greatest, are least able to address those needs.

Finally, local social movements have relatively little maneuverability in the current era of economic globalization and state restructuring. The extent to which they can exert any power whatsoever is a function of their own capacity set within the particular geographic and historical context in which they operate, especially the political opportunity structures that have evolved in place over time to empower nongoverning groups (Miller, 1994). The unique and changing relationship of the social movement to the local state is especially important, part of the larger process whereby civil society seeks to capture the state in serving its interests (see Mitchell, this volume and Wright, this volume). Furthermore, the history of conflictual politics between groups in a particular locale will play an important role in defining the specific response of each locality to economic globalization and state restructuring, and new conflicts—and coalitions—may arise in response to particular initiatives because power relations are most open to reconfiguration during periods of rapid political change. There are, nevertheless, limits to the power of local social movements in the present era. As devolution and privatization of the national state "subjects labor to global discipline, locally applied" (Peck, 1995, p. 256), rising job insecurity, declining wages and benefit packages, and the need to work longer hours limit the financial resources and time available to individuals who might contribute to such coalitions. Once again, the less affluent and politically experienced individuals will suffer the greatest risk.

Despite the limited power of states, organizations, and movements operating at the local level, Peck (1995) argues that solutions must be sought at higher levels:

> While appealing as a scale at which social mobilization can occur . . . local action can be only part of the solution and, uncoordinated, it may remain part of the problem . . . political action must confront the level at which the rules of regime competition are set (the supranational/global) not that at which the rules are carried out (the local). (p. 257)

Conceptualizing place in terms of its vertical linkages to other scales and its horizontal relations with other places, presented above, suggests strategies for augmenting the capacity to effectively address local needs. Local states, organizations, and groups can assert greater control by reaching up to those with kindred interests at other scales and by reaching out to those with similar concerns in other places (Cox & Mair, 1991; Miller, 1994; Smith, 1993). The *flexible use of scale and scope* draws in extralocal resources and expertise, expanding the local power to confront "the level at which the rules are made" (Leitner & Silvern, in press).

There are many successful examples of these cross-scale and translocal alliances. The traditional model is perhaps best illustrated by the United Way, a confederation of local nonprofit organizations each raising funds from private firms and individuals and then distributing these funds to other nonprofit agencies that address health, housing, and family services in the community. Local United Way organizations are coordinated at the national level by United Way of America, which provides support services to these local efforts (Karger & Stoesz, 1994).

Recent research has begun to document the processes whereby local social movements creatively use scale to build power, flexibly defining the arena for political action and directing their efforts to that level most open to change. For example, Herod (in press) examines the postwar history of the International Longshoremen's Association, representing East Coast waterfront workers. In response to political and economic changes in the shipping industry, the ILA devised an explicitly geographic strategy, forcing a change in bargaining practices from port-specific agreements to a master contract that held for a time at the national level. Herod argues that struggles over the scale at which groups negotiate have very real consequences, in this case rewriting the economic landscape of the industry, "a workers' applied geography." In a related study, Miller (1994) examines the recent history of the peace movement in Cambridge, Massachusetts, which flexibly directed its efforts to different scales of negotiation according to the specific issue involved and the particular political opportunities available at the local and central-state levels, whichever appeared most favorable to its interests at a given time. The local group then spread the results of its actions in a massive outreach project that affected the political mobilization of local peace movements in more than a hundred cities nationwide. Cross-scale and trans-local alliances such as these have begun to appear in local states, nonprofit organizations, and social movements spanning the political spectrum as each seeks to augment its power. The combined processes of economic globalization and state restructuring stack the odds against these local entities, but their ability to prevail is place- and time-specific (Miller, 1994).

■ Conclusion

In this chapter, I have reviewed the primary strategies for state restructuring in American federal history and have considered the consequences of the current round of devolution and privatization for localities seeking to compete in an intensifying international arena. At the core of this ongoing search for the appropriate balance of power is the issue of geographically distributing authority and resources within the federal hierarchy, which responds to the diverse map of regional conditions and political cultures, yet protects national norms valued in a common American political culture.

Although most Americans would agree in principle that one's health, affluence, and well-being should not depend on one's zip code (Green, 1989, p. 460), there exists little overall consensus as to how the responsibilities of the state should be divided. Indeed, few Americans seem to appreciate how current changes in the scale and scope of the American state will affect their lives, much less how this effect will depend on the place where they live. Above all, the American people need to "rediscover government" (DiIulio & Kettl, 1995, p. 65), separating the actual successes and failures of the state from the barrage of rhetoric that obscures its role. Beyond that, political elites and citizens need to recognize that the implications of state restructuring will differ considerably from place to place across the country, reflecting spatial variations in the resources and capacities of public and private institutions to manage these changes, and given the great geographic complexity of the American people whose lives will be altered by the outcome. These patterns and trends deserve careful study; and their analysis and documentation should guide the difficult decisions that are being made not only in Washington, DC, but also in city halls and town meetings across the country.

REFERENCES

Agnew, J. (1987). *The United States in the world-economy: A regional geography.* Cambridge, UK: Cambridge University Press.

Bennett, R. (Ed.). (1990). *Decentralization, local governments, and markets.* Oxford, UK: Oxford University Press.

Beynon, H., & Hudson, R. (1993). Place and space in contemporary Europe: Some lessons and reflections. *Antipode, 25,* 177-190.

Clark, G., & Dear, M. (1984). *State apparatus: Structures and language of legitimacy.* Boston: Allen and Unwin.

Cockburn, C. (1977). *The local state.* London: Pluto Press.

Cox, K. (1991). Questions of abstraction in studies in the new urban politics. *Journal of Urban Affairs, 13,* 267-280.

Cox, K., & Mair, A. (1988). Locality and community in the politics of local economic development. *Annals, Association of American Geographers, 78,* 307-325.

Cox, K., & Mair, A. (1991). From localised social structures to localities as agents. *Environment and Planning, A, 23,* 197-213.

DiIulio, J., & Kettl, D. (1995). *Fine print: The Contract with America, devolution, and the administrative realities of American federalism.* CPM Report 95-1. Washington, DC: Brookings Institution.

General Accounting Office (GAO). (1991). *Government contractors: Are service contractors performing inherently governmental functions?* GGD-92-11. Washington, DC: Government Printing Office.

Green, M. (1989). State attorneys general move in: Filling the deregulatory vacuum. *Nation, 249,* 441, 458-460.

Greenberg, M., Popper, F., & West, B. (1991). The fiscal pit and the federalist pendulum: Explaining differences between states in protecting health and the environment. *Environmentalist, 11,* 95-104.

Harvey, D. (1982). *The limits to capital.* Oxford, UK: Basil Blackwell.

Harvey, D. (1985). *The urbanization of capital.* Baltimore, MD: Johns Hopkins University Press.

Harvey, D. (1989). *The urban experience.* Oxford, UK: Basil Blackwell.
Herod, A. (in press). Labor's spatial praxis and the geography of contract bargaining in the U.S. East Coast longshore industry, 1953-1989. In H. Leitner & S. Silvern (Special Issue Eds.), *Political Geography, 15.*
Jessop, B. (1993). Towards a Schumpeterian workfare state? Preliminary remarks on post-Fordist political economy. *Studies in Political Economy, 40,* 7-39.
Karger, H., & Stoesz, D. (1994). *American social welfare policy: A pluralist approach.* White Plains, NY: Longman.
Kodras, J. (1990). Economic restructuring, shifting public attitudes, and program revision: The politics underlying geographic disparities in the Food Stamp Program. In J. Kodras & J. P. Jones III (Eds.), *Geographic dimensions of U.S. social policy* (pp. 218-236). London: Edward Arnold.
Le Grand, J., & Robinson, R. (Eds.). (1984). *Privatization and the welfare state.* London: Allen and Unwin.
Leitner, H., & Silvern, S. (Special Issue Eds.). (in press). *Political Geography, 15.*
Malbin, M. (Ed.). (1996). Symposium: American federalism today. *Rockefeller Institute Bulletin.* New York: Nelson A. Rockefeller Institute of Government.
Massey, D. (1994). *Space, place, and gender.* Minneapolis: University of Minnesota Press.
Miller, B. (1994). Political empowerment, local-central state relations, and geographically shifting political opportunity structures. *Political Geography, 13,* 393-406.
Mouffe, C. (1995). Post-Marxism: Democracy and identity. *Environment and Planning, D: Society and Space, 13,* 259-265.
Nathan, R. (1996). The devolution revolution. *Rockefeller Institute Bulletin.* New York: Nelson A. Rockefeller Institute of Government.
Offe, C. (1985). *Disorganized capitalism: Contemporary transformations of work and politics.* Cambridge, MA: Polity.
Osborne, D. (1993). A new federal contract: Sorting out Washington's proper role. In W. Marshall & M. Schram (Eds.), *Mandate for change* (250-261). New York: Berkley.
Pagano, M., & Bowman, A. O'M. (1995). The state of American federalism, 1994-1995. *Publius: The Journal of Federalism, 24,* 1-21.
Peck, J. (1995). *Work-place: The social regulation of labor markets.* New York: Guilford.
Peck, J. (1996, April). *Permeable welfare? Workfare politics and the deconstruction of Canada's work-welfare regime.* Paper presented at the Crises of Global Regulation and Governance Conference, Athens, GA.
Peck, J., & Tickell, A. (1994). Searching for a new institutional fix: The *after-*Fordist crisis and the global-local disorder. In A. Amin (Ed.), *Post-Fordism: A reader* (280-315). Oxford, UK: Basil Blackwell.
Peterson, P. E. (1995). *The price of federalism.* Washington, DC: Brookings Institution.
Pines, B. (1982). *Back to basics: The traditionalist movement that is sweeping grassroots America.* New York: William Morrow.
Pressman, J., & Wildavsky, A. (1973). *Implementation: How great expectations in Washington are dashed in Oakland; Or, Why it's amazing that federal programs work at all, This being a saga of the Economic Development Administration as told by two sympathetic observers who seek to build morals on a foundation of ruined hopes.* Berkeley: University of California Press.
Robinson, K. (1961). Sixty years of federation in Australia. *Geographical Review, 51,* 1-20.

Salamon, L., Musselwhite, J., & de Vita, C. (1986). Partners in public service: Government and the nonprofit sector in the welfare state. In *Philanthropy, voluntary action, and the public good* (42-58). Washington, DC: Independent Sector Inc.

Schwab, R. (1988). Environmental federalism. *Resources, 88,* 6-9.

Smith, C. (1986). Equity in the distribution of health and welfare services: Can we rely on the state to reverse the "inverse care law"? *Social Science and Medicine, 23,* 1067-1078.

Smith, N. (1993). Homeless/Global. Scaling places. In J. Bird, B. Curtis, T. Putnam, G. Robertson, & L. Tucker (Eds.), *Mapping the futures: Local culture, global change* (pp. 87-119). London: Routledge & Kegan Paul.

Storper, M., & Walker, R. (1989). *The capitalist imperative: Territory, technology and industrial growth.* Oxford, UK: Basil Blackwell.

Swyngedouw, E. (1992). The Mammon quest: "Glocalization," interspatial competition, and the monetary order: The construction of new spatial scales. In M. Dunford & G. Kafkalas (Eds.), *Cities and regions in the New Europe: The global-local interplay and spatial development strategies* (39-67). London: Belhaven.

U.S. House of Representatives, Committee on the Budget. (1991). *Management reform: A top priority for the federal executive branch.* Serial #CP-4. Washington, DC: Government Printing Office.

Warf, B. (1996, April). *The hypermobility of capital and the collapse of the Keynesian state.* Paper presented at the Crises of Global Regulation and Governance Conference, Athens, GA.

Wolch, J. (1990). *The shadow state: Government and voluntary sector in transition.* New York: Foundation Center.

Wolpert, J. (1993). *Patterns of generosity in America: Who's holding the safety net?* New York: Twentieth Century Fund Press.

6

How Federal Cutbacks Affect the Charitable Sector[1]

JULIAN WOLPERT

A persistent component of conservative ideology in the *Contract With America* and subsequent legislative debates is the view that charity has been suppressed by the welfare state and should be allowed to reassume a major remedial role for America's troubled families. The proposals anticipate that private contributions will expand to fill safety-net gaps and that charitable organizations will take the lead in reforming the flawed welfare system with a more caring and interventionist approach (Gillespie & Schellhas, 1994).

Greater reliance on charitable agencies can have a downside as well. The evidence presented in this chapter will show that their revenue sources are limited, highly dependent on government, unevenly distributed across the country, already highly targeted to existing service programs, and vulnerable to economic swings. Moreover, the severe brand of tough love and paternalism advocated by conservatives is reminiscent of 19th- and early 20th-century experiments in Social Darwinism whose failures helped provide the rationale for post-Depression entitlements and the emergence of modern welfare programs.

This chapter consists of four sections. I first identify a number of questions that Americans must address in order to understand the nonprofit sector and its ability to satisfy the demands expected of it. I then seek to answer these questions, based on a review of recent empirical research on charitable nonprofit organizations. The second section identifies the general characteristics of the nonprofit sector, including a description of the role charitable organizations currently do and do not perform in society and a portrait of recent geographic patterns and growth trends in the sector. In the third section, I investigate the capacity of the nonprofit sector to take on additional societal responsibilities, addressing in detail for this particular sector the general issues of institutional capacity raised in the previous chapter (see Kodras, this volume). Then, in the fourth section, I explore the changing relations between the public, private, and nonprofit sectors in the current era of government restructuring, identifying how proposed federal cutbacks will affect the capacity of nonprofits to take on responsibilities shed by government.

These issues can be addressed only in a preliminary fashion in this brief chapter. Fortunately, improved information is becoming available to examine tendencies and shifts in the charitable attitudes and donor behavior of the American population. Data resources make it possible to provide some estimates of the current capacity of charitable organizations, the extent of present unmet service needs, and the prospects for these organizations to enhance their revenue base with or without government aid. In particular, I will draw on a recent survey of charitable nonprofits conducted by the Independent Sector (Hodgkinson, Weitzman, Noga, & Gorski, 1995) and my recent research on the capacity of the charitable sector (Wolpert, 1993, 1995, 1996).

■ Questions Americans Need to Ask About the Nonprofit Sector

What can Americans really expect from philanthropy and charitable organizations? What is the proper place of these institutions in American society? Can and should the charitable sector do more to substitute for government? The answers to these and similar questions lie somewhere between the ambitious claims heard in the halls of Congress and the status quo. Government and the charitable sector had evolved, since the Great Depression, a rather stable partnership and division of responsibility, until challenged initially during the Reagan and Bush administrations and to an increasing extent since 1995.

Most Americans, and even a significant number in Congress, are poorly informed about the charitable sector in the United States—what it is, what it does, its current and potential capacity, and how donors and government can possibly enhance the latent capacity of nonprofit organizations. The challenge for charitable organizations to do more raises some fundamental questions about American generosity. Can donations and voluntary efforts be augmented—and at what price? Are Americans prepared to give a larger share of their income and wealth to support charitable organizations? Are they prepared to spend more hours as board members and fund raisers, and in training as volunteers for front-line provision of services? Are they willing to extend this added assistance across their community, age, ethnic, racial, and class boundaries to individuals and groups that most need their help? Do Americans believe that the needy have a legitimate and meritorious claim that deserves additional commitment? What is the proper role for charitable organizations or for government in providing assistance? Should assistance be provided on a temporary basis in emergencies only, or is the need for assistance long-term? Are there limits to "altruistic" giving—that is, does additional giving beyond token gifts require something in return? Is there evidence that donor and volunteer fatigue are already present?

■ Characteristics of the Nonprofit Sector

Nonprofit activities include *philanthropy* (enhancement of civic institutions), *charity* (assistance to the needy), and *service* (provision and delivery of services). Some nonprofit activities are solely charitable or philanthropic or service-oriented, but most reflect some combination of these three objectives. *Philanthropic activities* (generally by foundations and large donors) target the establishment and enhancement of institutions such as hospitals, universities, museums, and community social capital to work toward solving significant social problems. *Charitable organizations* (e.g., Salvation Army, Catholic Charities) transfer resources from the more fortunate to the needy. Nonprofit *service activities* aim to foster mutual benefit and pluralism, serve "thin" markets not served by private or public sectors, and enhance quality, variety, compassion, and efficiency in service delivery.

Nonprofit charitable and service organizations raise their own revenues from donations, government grants, and contracts and from dues, fees, and service charges. These revenues, in conjunction with tax benefits, enable them to provide services without charge or at reduced rates. Sliding fee schedules also enable some nonprofits to cross-subsidize service provision to their needier clients. Many nonprofits are also active on behalf of their clients in advocacy and lobbying efforts to improve social legislation and funding (O'Neill, 1989).

Charitable nonprofits now account for about 7% of national income and about 6% of total U.S. employment (Hodgkinson, et al., 1992). Their revenues and expenditures are considerable ($350 billion), but they are only about one-seventh the combined level of federal, state, and local government spending in these program areas. As recently as the mid-1950s, charitable organizations raised 70% of their income from donations. Now, only about 9% of their revenues is derived from individual, corporate, and foundation gifts and donations, compared with 37% from government grants and contracts and 54% from dues, fees, and other charges. The share of total revenues provided by government varies by sector (36% in health services, 17% in education, 42% in social and legal services, and 11% in the arts). Despite this variability, a significant component of nonprofit capacity is *the revenue that comes from government.* These funds are especially critical in social and health services that are targeted to low-income people. Diminution or withdrawal of government support implies a significant reduction in the capacity of nonprofits to assist low-income clients.

What Charitable Agencies Presently Do

Nonprofits are the major front-line service providers in their communities, even if their funding is partially derived from government grants and subsidies. Through donations and the efforts of volunteers, nonprofits reduce the financial burden on local government. Nonprofits spend money in their own communities. They contribute to the local economy and pay rent, wages, and wage taxes. They are the providers of community amenities like museums, symphony orchestras, and ballet companies that add to residents' quality and variety of life and that even enhance neighboring property values and revenues from tourism.

Nonprofits employ women and minority community residents to a greater extent than does the private sector, albeit at somewhat lower salary and benefit levels. Nonprofits enhance upward job mobility of their staff, volunteers, and clients through training for private and public sector employment. Volunteers are drawn from the local community, and community board members are often influential in public affairs and business and thus able to enhance their organizations' operations and missions.

Charitable organizations have the infrastructure (i.e., facilities, management, trained staff, and networks) to do more than they currently do. The great strengths of locally based organizations are that they tend to reflect local secular or religious values and that their services often represent extensions of family and neighborhood networks. Most are small scale and nonbureaucratic. They are often able to tap the local civic and altruistic motives of donors, volunteers, and businesses. They are often able to provide more personal and caring assistance to clients than is the case for government agencies or private-sector service firms.

On the other hand, charitable organizations are underrepresented in many neighborhoods and communities. Some agencies need to improve their management and enhance their administrative capacity. Some allow excessive and dysfunctional donor control to prevail over service provision. Targeting of services can sometimes be problematic and out of synch with community needs. Recipients may be subjected to paternalistic, moralistic, and conformity requirements as a condition for accepting assistance.

What Charitable Agencies Do Not Presently Do

Most donations that charities raise are needed to support community churches and synagogues, YMCAs, museums, public radio and TV stations, universities, and parochial schools—services that donors themselves use—and are not freely available to target the neediest and to sustain safety nets. The current pattern of donations does not transfer substantial cash payments to low-income people or sustain mentally and physically handicapped people, except for pure maintenance needs during local emergencies (Ostrander, 1989). The type of direct and personal bonding between individual donors and the needy that a number of politicians currently advocate is probably not feasible in the short run, even if its virtues over current practices could be established. Furthermore, the intensity of personal attention would be too costly if provided by nonprofit professional staff. Voluntary activity is certainly not at this level now, nor could it likely be augmented to such a magnitude to cope with even current needs.

Nonprofits cannot easily transfer their resources and aid from communities of affluence to places of need, except in the case of large national charities specializing in relief efforts. Because donors generally target their contributions for services in their own communities, funds raised in affluent suburbs cannot easily be retargeted to center cities or rural areas where resources are more limited and needs are greater. Thus, only about 10% of charitable contributions are targeted to the poor, and most charities lack the mechanisms to reallocate donations where they are needed most (Clotfelter, 1992). Locally based charities in most states also lack the organizational structure to lobby effectively with state and local government for more comprehensive client aid as the nationally based organizations have learned to do at the federal level.

Nonprofits are pervasive in center cities, suburban areas, and many rural communities throughout the nation. Although some organizations, such as the leading disease campaigns, are national in scope, most are rooted in communities both for fundraising and service provision. Locally based nonprofits are represented in all service sectors, including educational, social, health, cultural, and religious services, and are available to all income, ethnic, and racial groups.

Recent Geographic Patterns and Trends

The data show that the 85 largest metropolitan areas containing about 55% of the U.S. population were the home of about 220,000 nonprofits (501(c)3s) in 1989, about half of which were active service providers with annual revenues of more than $25,000 (Wolpert, 1996). The center cities now comprise only 39% of the metro populations, but contain two thirds of the nonprofits. Across the country, net growth (births minus deaths) of the nonprofits was more than 16% in the two years between 1987 and 1989, faster than the 2.2% population growth and faster than the 15% job growth in that period. The numbers of nonprofits grew faster in the suburbs (19%) than in the center cities (14.7%). Remarkably, the number of nonprofits grew by almost 15% in the center cities despite their near-stagnant 1% population growth and only 2.4% employment growth.

Center cities and suburbs also differ in the presence and growth rates of specific types of nonprofits. The center cities have greater shares of arts and cultural institutions and health facilities, whereas educational and human services organizations have a larger share in the suburbs. However, all the service categories grew faster in the suburbs between 1987 and 1989. Among center-city institutions, human service agencies grew only modestly faster (17.4%) than the other sectors. When one considers the enormous fund-raising challenge to both sustain center-city institutions and at the same time create a whole new suburban infrastructure, the enormity of the dislocation effects becomes readily apparent.

The aggregate trends in nonprofit representation for all the metro areas do not adequately convey the considerable variation between them and between their center cities and suburbs. The numbers of nonprofits grew consistently with population growth and increases in minority population. Nonprofits also increased more (but not beyond the significance threshold) in the wealthier metro areas and in those with smaller proportions of population in poverty. The variations in metro-area growth rates also reveal, although imperfectly, the catch-up or lag in nonprofit representation that has been taking place as a result of population shifts. The density of nonprofits (i.e., the number of organizations per capita) is still considerably larger in center cities and in the stagnant or declining metro areas. Assuming an equilibrium tendency toward a more uniform distribution of nonprofits, one might expect even greater nonprofit births in growing suburbs and metro areas and slower or no net growth in the stagnant zones.

Findings for America's 85 largest metropolitan areas show that nonprofits

- have been growing in numbers much more rapidly than the population;
- have been increasing more rapidly in the faster growing areas of the country than in stagnant or declining zones;
- are growing more rapidly in the suburban portions of metropolitan areas than in center cities;
- are increasing even in center cities experiencing substantial population and employment declines; and
- are increasing rapidly where minority populations are large.

■ The Capacity of the Nonprofit Sector

What is the *capacity* of the charitable nonprofit sector? Do the charitable organizations have the resources and the infrastructure to do more? Capacity refers to the capability of the sector to provide services and carry on its other missions (including public education, advocacy, etc.). The sector's capacity is a function of its resources (i.e., assets, income, personnel, management, volunteers, governing boards, and goodwill). The sector's latent or potential capacity is the gain that could be achieved through improved use of current resources and with the additional resources that could likely be harnessed in the near future.

As a rough guide to assessing capacity, I will focus on the charitable nonprofits (i.e., 501(c)3s) that are independent (unaffiliated) and report annual receipts exceeding $25,000. These currently number about 125,000 organizations out of a total one million nonprofits (Bowen et al., 1994). It is the capacity and activities of these larger and independent organizations, especially those in the human service sector, that are of primary concern here.

How Much Charity and Philanthropy Are Needed?

Efforts to encourage charity, philanthropy, and voluntarism within the American population are difficult and challenging; and as a society we rarely address the more abstract and normative issues, such as how much money should be raised, what share should be spent locally and for which kinds of services, and which functions should be left to the public or private sectors to perform. Answers to these questions require a community-based assessment of charity, philanthropy, and the nonprofit sector that can help us to understand donor behavior, as well as the magnitude and variety of the charitable services that the donations make possible.

Assessments of community quality of life are generally based on the level and variety of available services and amenities. Knowledge of the balance and range of service institutions and facilities in a community thus allows us to assess its nonprofit and public sector performance as well as its service deficiencies. We can also begin to assess the consequences of these deficiencies for the local community and to estimate whether the service gaps can best be addressed through private donations or through public sector expenditures.

Donation levels should be of sufficient magnitude to enable a community to support a complement of services consistent with its size, affluence, distress, compassion, civic obligations, preferences, and tastes. The service complement of a metropolitan area is predictable in general terms. For example, we know the expected population level needed to support a range of social and health services, such as an art museum, a ballet company, or a public TV station. We also know how the likelihood of having these institutions represented in the community is affected by social and cultural preferences and local economic well-being in the locale. On the one hand, high levels of community generosity should translate into meeting or even exceeding the threshold or service complement. On the other hand, stinginess would likely translate into service gaps.

Private giving and public sector support have each evolved distinctive service niches. If the generosity consists primarily of charitable donations and philanthropy, then one would expect that the community will have elaborate religious institutions with better paid clergy, fine museums with more extensive collections, strongly supported private schools, better equipped hospitals, and more comprehensive human services. If the generosity is in the form of greater local tax effort channeled primarily through the public sector, then its effect should be reflected in well-supported public colleges and universities, better public libraries with higher paid librarians, more recreational and open space, and more substantial income transfers to the needy and handicapped.

How Much Do Americans Currently Contribute?

Contributions to charitable nonprofits currently total almost $130 billion, of which about 87% comes from individual donations and bequests, somewhat above 7% from foundations, and somewhat less than 5% from corporate gifts (AAFRC Trust for Philanthropy, 1994). Forty-five percent of the donations were targeted to religious institutions; 12.9% to education; 8.9% to health and hospitals; 9% for human services; and 7.5% for arts and culture.

American households, including noncontributors, donated an average of $646 or 1.7% of household income in 1994, according to the most recent Independent Sector survey (Hodgkinson et al., 1994). Nearly 75% of all households reported some contributions. Among *contributors,* donation levels averaged 2.1% of household income. The giving level among *all households* was 1.5% in 1987, 2.0% in 1989, 1.7% in 1991, and 1.7% in 1993—and thus down by 6% in constant dollars from 1991 and 1993 and down by 19% from 1989 to 1991 after an increase during the 1987 to 1989 period. The changes reflect lower giving rates, especially by corporate executives and their firms during recessionary periods.

Findings from the Independent Sector survey indicate that giving can be increased somewhat if more are asked to contribute, volunteer, and become members of charitable associations and if more had access to payroll deduction plans for contributions at their place of work (Hodgkinson et al., 1994). Charitable organizations are already making extensive use of such innovations as charitable remainder trusts and gift annuities, life estate agreements, donor advised funds, and gifts of real estate and other property.

About 90% of present charitable contributions are both raised and spent locally. Some communities (such as Minneapolis, Pittsburgh, and Cleveland) are much more giving than others in both their private and public sector generosity, even after allowance is made for differences in affluence and need (Wolpert, 1993). Very little is redistributed between rich and poor communities or sent abroad, except for short periods when domestic or foreign crises are especially severe and well publicized. Lacking a more inclusive notion of community within American society (see Staeheli, this volume), philanthropy cannot be expected to have more than a limited and highly selective role as a supplement to the public sector. Certainly, philanthropy and charitable behavior perform valuable functions, but we should not assume that these can be redirected to cover for a reduced role by government.

My recent research assessing America's generosity in the charitable sector (Wolpert, 1993) shows that

- the American population generally accepts the partnership that has evolved between government and the charitable nonprofits and does not readily distinguish between the generosity of the public and charitable sectors;
- the growth of charitable nonprofits in the 1980s and early 1990s can largely be attributed to their greater revenues from federal, state, and local government grants and contracts and from higher client fees for their services;
- charitable organizations have neither the current nor the potential capacity to assume a significantly larger role in providing and delivering services without more government aid (i.e., they cannot do more with less); and
- charitable donations to nonprofits have hit a ceiling that cannot be increased without greater prosperity in the nation (i.e., when the need for charity is diminished), larger tax incentives (which are unlikely in the current budget-balancing mood), and more widespread use of donor choice in funding allocation.

Of course, the agenda of those who urge a greater role for charitable organizations may be fiscal as well as moral. The proposed federal cutbacks and recisions will have a selective impact on services to the poor and minorities. In fact, the evidence presented in my earlier study (Wolpert, 1993) showed that the transfer of government services to charitable organizations would likely yield highly uneven returns:

- Americans are not uniformly generous but vary significantly from place to place in the amounts they give and the type of charitable organizations they target;
- geographic disparities in generosity levels have been declining, but this is principally due to a harsher economic environment in the more generous places rather than to greater generosity in the more parsimonious places;
- generosity is greater in places where per capita income is higher and increasing, in areas where the political and cultural ideology is liberal rather than conservative, and in the smaller metropolitan areas where distress levels are lower;
- areas demonstrating greater generosity show higher levels of giving targeted to educational, cultural, and health services benefiting donors and lower levels of assistance providing a safety net to the needy;
- nonprofit organizations vary substantially in the level of financial resources and therefore in their ability to provide basic services to the lowest income population;
- the increased sorting of Americans into socially homogeneous communities has targeted public and nonprofit services in the suburbs where affluent donors/taxpayers reside, often at the expense of support for center-city and rural institutions;
- gaps in the provision of basic services and the general quality of life are widening between the growing and declining states and within metropolitan areas; and
- our severely fragmented and atomized nonprofit sector contributes effectively to the variety and quality of life in American communities but lacks the resources and institutional structure to provide a safety net for the needy or to address regional disparities.

Recent Social/Demographic Shifts and Nonprofit Sector Capacity

America's nonprofits have been profoundly affected by massive demographic and social change affecting the U.S. population over the past several decades. Among America's largest metropolitan areas, the center cities still contain most of the metro nonprofit organizations but suffer from declining shares of population and jobs, as well as eroding suburban interest in maintaining center city institutions. Compounding the nationwide suburbanization trend has been a considerable shift of people and jobs away from the Northeast and Midwest regions, the traditional heartland of nonprofit institutions and generous contributions.

Poverty has increased significantly, especially in those center cities that already had large representations of low-income African-American and Latino residents or were entry ports for recent immigrants. These demographic and social changes have drastically altered service needs in center cities and the activities of nonprofit service providers. Those nonprofits founded to provide social, educational, and cultural services to middle and working-class whites are either readjusting to new client groups (albeit, with more government assistance), ceasing their operations, or relocating to the suburbs. Critical management decisions by nonprofits and their philanthropic supporters will be necessary over the next several years to help the sector adjust to these shifts.

In the center cities of many metropolitan areas, nonprofit organizations in the health and human service, educational, and cultural sectors are the major service providers and employers. However, their sources of support have altered drastically. They have become increasingly dependent on government grants and contracts as fees for service and individual contributions account for decreasing shares of their budgets. The shifts in support have left these nonprofits (and their service clients and staff) highly vulnerable to uncertainties in federal support programs and pass-throughs and to the constraints of state and local finances. Further federal cuts and the privatization of programs are likely to have major impacts on center-city nonprofits at a time when they are likely to face increasing service demands and a more difficult fund-raising environment for private donations.

Private service firms have greater flexibility in responding to urban and regional changes than is the case for the public sector and nonprofits. Private firms can close offices in declining center-city neighborhoods or shift their facilities and investments to the suburbs, other regions, or metro areas in response to market shifts. In contrast, nonprofits are more grounded at particular sites, facilities, and locations by their charter, service mission, support base, and relationship to trustees. Nonprofits cannot so readily shift their locational base, relocate their facilities, or alter their pattern of service provision in response to market conditions, neighborhood social change, suburbanization of residents or jobs, or interregional shifts in economic growth. Nonprofits' greater attachment to place can be a decided benefit for the remaining service clients in declining neighborhoods and cities, but only if they can maintain sources of financial support. The greater flexibility that private sector firms have in responding to client and market shifts provides some potential lessons for nonprofits that merit assessment and evaluation. We need to document the financial pressures, constraints, and opportunities that nonprofits face and find ways to help them continue their important community and neighborhood roles.

The financial viability of a community's nonprofit sector is closely related to its region's economic vitality. The most significant revenue streams for nonprofit organizations depend more on local wealth, wage levels, and direct donor benefits from nonprofit services than on local need or levels of distress. Thus, nonprofits make a significant and growing contribution to a region's economy, but the economy contributes instrumentally to the well-being of nonprofits as well.

■ Changing Relations Between Public, Private, and Nonprofit Services

The current era of government restructuring is fundamentally rearranging the relationships between the public, private, and nonprofit sectors as the federal government seeks to transfer responsibilities to lower tiers in the governmental hierarchy, to private corporations, and to nonprofit organizations. I explore the implications for the nonprofit sector below.

Could an Expanded Role by Charitable Organizations Be Harmful?

An increased role by charitable organizations may intrude into realms more effectively handled by private business or usurp societal responsibilities vested in government within advanced and affluent democratic societies. Charitable fundraising to supplement public funding of public education, libraries, and open space preservation, for example, implies abdication of public responsibility to provide these services at a satisfactory level. Greater charitable donations to help the poor similarly implies surrender of equity ideals by an affluent society in public sector life.

In addition, nonprofits pay no property or corporate taxes and have been accused of unfairly competing with private sector firms. Nonprofits sometimes "cream" clients from private service firms and "dump" needy clients on public sector services. Nonprofit boards are often not representative of community diversity. Some nonprofits are suburbanizing at a fast pace and are siphoning off philanthropic investments for new buildings and facilities, often at the expense of maintaining center-city service institutions. Furthermore, nonprofits are rarely able to provide services in rural communities. The numbers of nonprofits are proliferating in some communities, and many resist necessary readjustment to improve efficiency. They consume funding in salaries and rents and other administrative and fund-raising costs that would otherwise be available for providing services.

Many nonprofits have lost some of their independence and are increasingly agents of the state or their corporate and foundation sponsors (Smith & Lipsky, 1993). Funding needs have made them dependent on the government's agenda, and dependence on corporate support increasingly requires pursuing a utilitarian agenda. Dependence on foundation grants often necessitates a stream of start-up efforts in innovative new programming, sometimes to the detriment of ongoing service programs. Nonprofits often neglect service evaluation and attention to client satisfaction because performance is more frequently assessed by inputs rather than outcomes for clients. An expanded role for the charitable sector would need to address all of these concerns.

Finally, problems arise if the nonprofit sector grows faster than the public or private sectors. Salary and wage levels are substantially lower in the nonprofit sector than those in the private market, so relatively faster growth in nonprofit employment would mean a relative decline in corporate and employee donations and user fees and reduced capability of the public sector to provide safety-net services.

Impacts of Federal Cuts on Charitable Nonprofits

In the 1980s, the challenge for charitable agencies to do more with less was met by protests and lobbying efforts that helped to neutralize threatened federal cutbacks. In fact, the decade passed with less overall diminution of government support for nonprofits than had been feared. However, the current cutback proposals are expected to have more severe impacts by simultaneously increasing the demand for nonprofit services while cutting back government contributions to nonprofits. American communities and their charitable agencies will face a severe test when the cuts and their multiplier effects are felt. Competition for donations is expected to increase sharply without any assurance that the charitable pie can be enlarged.

Nonprofits specializing in social services and legal services were the most seriously affected by cutbacks in the early 1980s when the federal government cut almost $40 billion in welfare, job training, and housing services (Salamon, 1987). Government contracts to nonprofits for such services as job training and maternal and child health dropped by $47 billion during the 1982-1992 period, and overall government support as a share of nonprofit revenues declined by 20% over the decade. A 7% increase in contributions and the 48% increase in payments from dues, fees, and service charges did not make up for this decline in public sector support.

According to a recent Independent Sector analysis of the 1995 House budget resolution, proposed federal cuts would have a severe impact on charitable organizations (Hodgkinson et al., 1995). Nonprofits would face a $254 billion cumulative gap in their revenues during the Fiscal Years 1996-2002 if the cuts are fully implemented. Proposals include a reduction of 25% in spending authority for job training programs; a 50% cut in the National Endowment for the Arts and Humanities and a 20% reduction in the Social Service Block Grant, among other reductions. Overall, the cuts in grants and contracts to nonprofit agencies would significantly diminish their capacity to deliver services unless donor contributions are able to make up the difference, a highly unrealistic scenario. These figures do not include the impact of welfare reform proposals that could make many recipients ineligible for welfare payments and more dependent on charitable agencies (see Cope, this volume). Medicaid cuts would also eliminate coverage for many children by the year 2002. According to Senator Daniel Patrick Moynihan, "There are ... not enough social workers, not enough nuns, not enough Salvation Army workers to care for children who would be purged from welfare roles were Congress to decree ... a two year limit for welfare eligibility." The proposed House cuts would reduce the federal share of charitable, nonprofit program spending by 25%, according to the Independent Sector's recent panel study (Hodgkinson et al., 1995). The sample of agencies included in the study consisted of 108 nonprofits in 31 states, representing all of the major service subsectors: family services; arts and culture; job training; services for youth, the elderly, and the handicapped; homeless shelters; higher education; and so on. More than three quarters of their present programs are at least partially financed by federal grants. They currently receive 32% of their revenues from government—the cuts would reduce this level to 25% by the year 2002.

The sample survey of nonprofits shows that cumulative charitable income would have to be augmented by nearly 70% from 1996 through 2002, over and above the real increases built into their projections. By

2002, these organizations would have to increase charitable contributions by 120% to offset the cumulative reductions in federal funding. The federal share of revenues for agencies serving the elderly would decline from 17% to 9% and require a 33% increase in donations. The federal share of revenues for housing and community development agencies would decline from 51% to 31% and require a 162% increase in donations. The federal share of revenues for agencies serving children would decline from 31% to 23% and require a 58% increase in donations. Most of the surveyed organizations ventured that they would try to increase their revenues from other sources, but they recognized that competition would be much stiffer. In the end, they expect to have to reduce staff and programs because earlier cuts have already absorbed whatever slack existed.

Limits to Increasing Donations and Voluntarism

Donations to nonprofits are not expected to increase substantially even if the proposed cuts are implemented. The 1993 real increase in donations was only 0.5% overall. Individual giving rose by 0.8% in 1993; the foundation increase was 0.9%; and corporate giving declined by 2.1%. Among the various service sectors, giving to social service organizations, such as Catholic Charities and the Salvation Army (which are among the most redistributive of nonprofit services), peaked in 1989 and had declined by 6% in 1993. There was also a real decline in arts giving in 1993. The largest increases were in the education sector (5.4%), health (3.3%), and religion (1.5%). The annual survey by *Chronicle of Philanthropy* of giving to the nation's 400 largest charities, which together account for about one sixth of annual giving, showed that donations rose by 6.3% in 1994. The Salvation Army raised more than in 1993, but the Red Cross and the United Way raised less.

The most optimistic estimates are that contributions might be able to make up for only 5% of federal cuts. Other estimates are as low as 1 to 2%. Even if giving levels were to rise, only a modest share of the increments would likely be targeted to assist the most disadvantaged groups. Higher giving rates typically benefit education, culture, and the arts rather than social services.

California's recession, for example, has severely affected giving to United Way, which relies on solicitation at the workplace. Donations in Los Angeles, which reached a high of $87.2 million in 1989-1990, dropped to only $57.2 million in 1994-1995, with corresponding declines in the funds allocated to local charities *(Chronicle of Philanthropy,* 1995). The Los Angeles United Way is attempting to reduce its overhead so that a greater share of contributions can be passed on to member agencies. These agencies, in turn, are trying to trim their overhead and some of their services while attempting to boost their own fund-raising efforts.

Finding ways to increase voluntarism is also problematic. Many Americans share with congressional opponents of welfare the same negative attitudes toward low-income populations and are thus unlikely to expend personal time and effort volunteering to help others.

Neither is it realistic to believe that the cuts in federal funding can be made up by charitable giving from corporations and foundations. For example, the total *assets* (not the income from endowments) of America's 34,000 foundations add up to only about 10% of current government expenditures for social welfare and related domestic programs. Foundation wealth is increasing rapidly, but most foundations do not currently even dispense the mandated yearly level of 5% of assets. Corporate giving is currently estimated at $6.2 billion or less than 1% of their pre-tax net income. Donations did rise substantially at an annual growth rate of 13% in the 1980s, but growth rates declined in the 1990s and are not likely to rebound even with more rapid national economic growth, because donation levels are lower among multinationals than among locally owned corporations.

What More Can Charitable Nonprofits Do?

Charities can attempt to increase their donations and volunteers and their contract funding from state and local government and from service fees charged to clients who can afford more. They can try to reduce their costs through greater efficiency, lower administrative overhead, reduced staffing, and lower salaries and benefits. Some efficiency gains may be realized through mergers of existing agencies. An active donors' forum, umbrella organization, or community trust board can help to allocate contributions to sectors where they are most needed in center cities and rural poverty areas. Cross-subsidies can be used more intensively.

Charitable organizations have become increasingly alarmed as details of the proposed recisions and cuts and the devolution of federal programs to states and localities become more specific and immediate *(Chronicle of Philanthropy,* 1995). The prevailing perspective, especially among the larger and more established agencies, is that the cuts will inevitably lead to curtailments primarily of those services for low-income clients that government has subsidized. A minority of agency heads argue that Americans will donate and volunteer more, that greater efficiency can be achieved in providing services, and that many existing nonprofit services supported by government should be eliminated or reduced in scope. The hazard, of course, is that restoration to even current service levels would be difficult if federal cutbacks are made, additional contributions are not forthcoming, and the nonprofit infrastructure is reduced or dismantled. Efficiency gains could and should be fostered in government and the nonprofit sector, but their impact on the overall situation should not be overestimated.

What More Can the Federal Government Do Within These Limits?

Government can encourage more charitable giving and volunteering by reducing the cost of giving and by allowing tax deductions for voluntary efforts. More specifically, government can restore the charitable deduction for both itemizers and non-itemizers and eliminate the 3% floor for charitable donations. In the balance of revenues lost through deductibility of charitable donations versus the gains from the enhanced capacity of nonprofits to provide additional services, the scales favor reducing the cost of giving through greater deductions. The major irony is that many who call for greater reliance on private charity also support flat-tax schemes that would end charitable deductions and thus suppress giving levels.

In fact, government has been making life more difficult for nonprofits. Federal, state, and local governments have increased their scrutiny of nonprofit organizations. They have been examining ways of restricting tax deductible gifts, subsidized nonprofit mailing privileges, involvement in revenue-producing services, fund-raising expenditures and practices, lobbying and advocacy, and exemption from property taxes. Notwithstanding the potential merits of this greater scrutiny, the general direction of current government involvement is likely to lead to lower nonprofit revenues and greater costs.

Furthermore, various forms of tax reform legislation now being proposed (e.g., reduction in graduated tax rates, a flat-rate tax, substitution of income taxes by a consumption tax, elimination of the capital gains tax, caps on deductible contributions, total elimination of charitable deductions) will raise the cost of giving and reduce giving levels. Tax incentives probably do not affect whether or not donors decide to make contributions, but they do influence the size of the gift by influencing the cost of giving. Estimates show that if there were no charitable deduction, donations would decline by one third, or by $20 billion, among the 32 million current itemizers. Evidence concerning the donations of the 81 million non-itemizers are quite clear as well. The Tax Reform Act of 1986, which eliminated the charitable deduction provision for taxpayers who use the standard deduction on their tax returns, reduced the growth of giving rates substantially, especially among middle-income households. The amounts contributed by itemizers are now more than twice the level given by non-itemizers in the same income categories.

More recent changes in federal tax law have further reduced the incentives for giving. The 1990 legislation placed a 3% floor on deductions (including charitable deductions) for taxpayers with adjusted gross incomes over $100,000. The floor has now been made permanent and has been adopted by 26 states in their own tax laws.

■ Conclusions

The current proposals to reduce the role of the federal government in service provision are not backed by evidence that either state and local governments or the organized charities can pick up the slack. Independent studies estimate that the recisions and cutbacks will greatly weaken the traditional support base and hinder access to health and social services by America's lowest income households. The Independent Sector's panel study quite effectively shows the extent of anticipated federal revenue losses to social service agencies and the very unlikely prospect that additional charitable contributions can make up even a small part of the difference.

Despite these dire predictions of lower safety-net protection and greater service gaps, will charitable donors simply stand by in the face of so much additional need? The evidence about charitable giving presented here shows that even if individual, foundation, and corporate donors were to retarget virtually all their contributions to relief of immediate social needs, the total amount of giving would not make a noticeable dent in the shortfall resulting from the federal cuts. What would happen in the meantime to all the other institutions that need continuing support from charitable giving—churches, universities, museums, symphony orchestras, and overseas relief? The large national social agencies, such as the Salvation Army, Red Cross, and Catholic Charities, are themselves highly dependent on federal grants for the services they provide and do not expect that they can raise enough compensatory funding. The impact of the recisions and devolution/privatization will not be fully felt until the last few years of the budget-balancing agenda, because the cuts are planned to be cumulative through the year 2002. In the meantime, a good deal of damage will be done to our most vulnerable households.

NOTE

1. Shortened and revised from Wolpert (1996).

REFERENCES

AAFRC Trust for Philanthropy. (1994). *'94 giving USA*. New York: American Association of Fund-Raising Counsel.
Bowen, W. G., Nygren, T. I., Turner, S. E., & Duffy, E. A. (1994). *The charitable non-profits: An analysis of institutional dynamics and characteristics*. San Francisco: Jossey-Bass.
Chronicle of Philanthropy, Vol. 7. (1995, August 10). Marion, OH: Chronicle of Higher Education.
Clotfelter, C. T. (Ed.). (1992). *Who benefits from the nonprofit sector?* Chicago: University of Chicago Press.
Gillespie, E., & Schellhas, B. (Eds.). (1994). *Contract with America*. New York: Times Books.
Hodgkinson, V., Weitzman, M. S., Noga, S. M., & Gorski, H. (1992). *Nonprofit almanac: Dimensions of the independent sector, 1992-93*. San Francisco: Jossey-Bass.
Hodgkinson, V., Weitzman, M. S., Noga, S. M., & Gorski, H. (1994). *Giving and volunteering in the United States*. Washington, DC: Independent Sector.
Hodgkinson, V., Weitzman, M. S., Noga, S. M., & Gorski, H. (1995). *The impact of federal budget proposals on the activities of charitable organizations and the people they serve*. Washington, DC: Independent Sector.
O'Neill, M. (1989). *The third America: The emergence of the nonprofit sector in the United States*. San Francisco: Jossey-Bass.
Ostrander, S. (1989). The problem of poverty and why philanthropy neglects it. In V. A. Hodgkinson & R. W. Lyman (Eds.), *The future of the nonprofit sector* (pp. 219-236). San Francisco: Jossey-Bass.
Salamon, L. M. (1987). On market failure, voluntary failure, and third party government. *Journal of Voluntary Action Research, 16*, 29-49.
Smith, S. R., & Lipsky, M. (1993). *Nonprofits for hire: The welfare state in the age of contracting*. Cambridge, MA: Harvard University Press.

Wolpert, J. (1993). *Patterns of generosity in America: Who's holding the safety net?* New York: Twentieth Century Fund.
Wolpert, J. (1995, Fall). Delusions of Charity. *American Prospect,* No. 23, pp. 86-88.
Wolpert, J. (1996). *What charity can and cannot do.* New York: Twentieth Century Fund.

7

State Restructuring and the Importance of "Rights-Talk"

DON MITCHELL

A common theme running through the chapters of this volume is that the state plays an intricate and often contradictory role in its relations with capital and civil society. In the current round of global restructuring, the state promotes capital accumulation—through such mechanisms as granting charters to corporations, establishing tax policies, and negotiating trade agreements—because under existing circumstances, success in the global economy appears necessary to its financial sustenance. Yet at the same time, the state must seek legitimacy from civil society if it is to remain in power in a liberal democracy. As globalization of the U.S. economy damages increasing portions of the American population and thus threatens the legitimacy of the state, various social movements have arisen within civil society to confront capital through the state. For example, Wright (this volume) describes how segments of the New Right, arguing that jobs in the United States are threatened by the globalization of the American economy (and of labor), seek to restructure the state so as to protect their interests. Patrick Buchanan's 1996 presidential campaign, based in part on a platform of protectionism and exclusion, emerged as a symbol of this movement.

In this chapter, I focus on another segment of civil society, arguing that the Left must also enter the fray and begin the hard work of recapturing the state in *its* own interests. Specifically, I will highlight three crucial and interconnected battlegrounds for Left social movements struggling with the restructuring of the state and the unfettered globalization of capital. These are the struggle for "rights," struggles over the production of space, and struggles over the use and production of scale. All three battlegrounds suggest that contrary to much leftist theorizing, capturing and democratizing the state must remain a central goal of any social movement seeking to create a more just society. The state is simply too important to be left to the Right and to capital.

■ The Necessity of Rights

In a recent essay—published in *Harper's* at about the same time the Supreme Court announced a decision striking down Colorado's anti-gay Amendment 2 (*Romer v. Evans* [1996])—the philosopher Richard Rorty (1996) argues that a reliance on "rights-talk" as the basis of political movements is wrong-headed. "The difference between an appeal to end suffering and an appeal to rights," Rorty argues,

> is the difference between an appeal to fraternity, to fellow-feeling, to sympathetic concern, and an appeal to something that exists quite independently from anybody's feelings about anything—something that issues unconditional commands. Debate about the existence of such commands, and discussion of which rights exist and which do not, seems to me a philosophical blind alley, a pointless importation of legal discourse into politics, and a distraction from what is really needed. (p. 16)

For Rorty, what is needed is simple: empathy by the majority for the minority. Rorty takes his critique a step further. He argues that rights-talk by the Left has led it to focus too narrowly on "sadism"—on the willing beating down of others as a means of promoting one's own, perhaps also rather tenuous, standing. This has led those concerned with the rights of others to neglect an important aspect of the American political and cultural scene: selfishness. Rorty (1996) argues that "selfishness differs from sadism in being more realistic and more thoughtful, less a matter of one's own worth and more a matter of rational calculation" (pp. 16-17). For Rorty, protection of minorities against sadism has been the *raison d'être* of "cultural politicians of the academic left" since the successes of the Civil Rights Movement. But in the past 30 years, Rorty argues, selfishness has been the bigger problem in America: "You would not guess from listening to the . . . left that the power of the rich over the poor remains the most obvious, and potentially explosive, example of injustice in contemporary America" (1996, p. 17).

To Rorty, rights-talk accords not with elimination of the "most obvious" injustice—economic injustice—but rather with an overweening concern with cultural domination. Hence,

> The more we on the American left think that the study of psychoanalytical or sociological or philosophical theory will give us a better grasp on what is going on in our country, the less likely we are to speak in a language that will help bring about change in our society. The more we can speak a robust, concrete, and practical language—one that can be picked up and used by legislators and judges—the more use we will be. (1996, p. 18)

The language of rights is just not capable of providing this "robust, concrete, and practical language." In fact, for Rorty, it positively hinders it. "If *Bowers v. Hardwick* [a 1986 decision upholding anti-sodomy laws] is reversed," Rorty asserts,

> it will not be because a hitherto invisible right to sodomy has become manifest to the justices. It will be because the heterosexual majority has become more willing to concede that it has been tormenting homosexuals for no better reason than to give itself the sadistic pleasure of humiliating a group designated as inferior—designated as such for no better reason than to give another group a sense of superiority. (1996, p. 16)

For Rorty, then, rights-talk must be replaced with a fully charged moral discourse, a discourse designed not to protect, but to convince, not to create legal entitlements and restrictions, but to produce political realities.

Leftist critique of rights-talk did not originate with Rorty. More than a decade ago, for example, legal scholar Mark Tushnet (1984) attacked the reliance of the Left on rights, arguing that such a reliance opened the door not to liberation but to domination. Like Rorty, Tushnet based his argument on the sense that rights-talk distracted from what is really needed to make a more just, more humane world. Tushnet (1984) claimed that "there do seem to be substantial pragmatic reasons to think that abandoning the rhetoric of rights would be the better course to pursue for now" (p. 1394) if we want to build a society that provides and nurtures a good life for all—or more immediately if we simply want to reduce suffering in the world.

> People need food and shelter right now, and demanding that those needs be satisfied—whether or not satisfying them can today be persuasively characterized as enforcing a right—strikes me as more likely to succeed

than claiming that rights to food and shelter must be enforced. (Tushnet, 1984, p. 1394)

Tushnet grounds his critique of rights discourse in four arguments: (a) Rights suffer from instability—that is, they are not universal and abstract, but rather can only exist as products of particular political and social moments, and thus as those moments change, so too do rights; (b) Rights suffer from indeterminacy—that is, "the language of rights is so open and indeterminate that opposing parties can use the same language to express their positions" (Tushnet, 1984, p. 1371); (c) Rights suffer from reification—that is, treating real experiences as instances of simple exercises in abstract rights "mischaracterizes" (Tushnet, 1984, p. 1382) and devalues those experiences, eliminating what is most important from any social action—its political efficacy; and (d) Rights suffer from political disutility—that is, rights often protect privilege and domination instead of the oppressed and minorities, as when commercial speech or the ability of corporations to hire petition-signature gatherers is "guaranteed" by the First Amendment.

Tushnet stakes much of the moral force of his argument on this last point: "It is not just that rights-talk does not do much good. In the contemporary United States, it is positively harmful" (Tushnet, 1984, p. 1386). This is so, in part, because "[t]he contemporary rhetoric of rights speaks primarily to negative ones. By abstracting from real experiences and reifying the idea of rights, it creates a sphere of autonomy stripped from any social context and counterposes to it a sphere of social life stripped of any content" (Tushnet, 1984, pp. 1392-1393). And equally important, "the predominance of negative rights creates an ideological barrier to the extension of private rights in our culture" (Tushnet, 1984, p. 1393). Yet this is not just a matter of better promoting positive rights (the right *to* something, rather than the right to be protected *from* something); for, as already noted, a reliance on a struggle for rights would seem to distract from the real needs at hand (like food or shelter).

In the context of the radical restructuring of both capital and the state that has quickened since the structural crises of the early 1970s, arguments such as those by Rorty and Tushnet need careful attention; for they raise precisely the sorts of questions that ought to be addressed by those who would fight for a more just world in the face of the reordering of political, economic, and social life that these restructurings demand for their own sustenance. Rorty is correct: The most serious problems facing the United States (and the rest of the world) center around the "power of the rich over the poor," and that, "[a]s Karl Marx pointed out, the history of the modern age is the history of class warfare, and in America today, it is a war the rich are winning, the poor are losing, and the left, for the most part, is standing by" (Rorty, 1996, p. 15). So too is Tushnet correct when he remarks that world-wide, "[t]hings on the whole are terrible," and since the United States (or more accurately capital coordinated through the political and military might of the United States) has created "one of the great empires in world history . . . life in the metropolis goes on as well as it does only because the metropolis exploits the provinces. (Tushnet, 1984, p. 1402)

In such a world, talk of rights might seem rather secondary. There are far more important battles to be fought, battles that are at once moral (Rorty) and political (Tushnet).

Yet for all the strengths of their analyses, something is lacking—something that the current restructuring of state and capital makes so readily apparent that it is hard to see how analysts as perceptive as Rorty and Tushnet could so easily sidestep it. And that is simply this: At a time when the globalization of capital is aided and abetted at every step of the way by states organized through what Stuart Hall (1988) famously called (in the British case) "authoritative populism"; when, under the name of free trade and unfettered markets, capital is free to systematically crush any vestige of social life not yet under its sway, free to create a world in which the immiserization of the many so as to aggrandize the very, very few is packaged as inherently *just;* then those who seek to create a better world have few more powerful tools than precisely the language of rights, no matter how imperfect that language may be.[1]

Let me put that another way. There is a central contradiction that all social movements must face—a contradiction that must be faced squarely even as it is hard to see how it can be overcome. On the one hand, one of the greatest impediments to freedom, to a just social life, to the kind of moral world Rorty would like us to believe we could live in, is the state itself. Tushnet is correct in arguing that rights codified through the institutions of the state can be enormously destructive. Moreover, the state itself is so fully complicit with the program of capital that it seems hopelessly utopian to think that it could ever be extricated and turned into a force of liberation. That is precisely why so many on the Left are willing to abandon state-centered approaches to social change (including rights-talk) and substitute for them either the stern moralism Rorty advocates, the cultural politics Rorty critiques, or the reliance on extra-parliamentary, extra-judicial politics Tushnet proposes.

On the other hand, the state has proved itself—precisely through the institutionalization of rights—to be key protector *of* the weak. These protections have not been freely given; they have been won, wrested from the state and from those it really protects through the moralism, the direct action, and the cultural politics that have marked political history under capitalism—that is, through unceasing struggle. But importantly, these fragile victories, incomplete as they are, counterproductive as they may sometimes be, are themselves protected only through their institutionalization in the state. As Meghan Cope argues in this volume, for women and children, the state—the U.S. federal state at that—has been, in many ways, the best friend they have. The creation of a progressively democratic state (or even a first step toward the dream of seeing the state wither away) must itself, in good part, begin by *strengthening* the state. Put another way, the state is an essential player in contemporary capitalism and will remain so, no matter how much current political trends promote the *appearance* of its demise after Keynesianism (see Meszaros, 1995). To abandon the state to the forces of capital or to those so efficient in organizing authoritarian populism *through* the state is shortsighted in the extreme.

I have begun by highlighting the contradictory nature of rights and states under capitalism because I think it is vitally important to understand these contradictions if we are to develop means to control and regulate both the state and capital—that is, to democratize the state such that capital can be properly controlled. But I also think understanding these contradictions (and learning to work within them) is vital to understanding the nature of resources that social movements will have to reinforce and reinvent in the era ushered in by the globalization that attended the crises of the 1970s and the new worlds created out of them. I want to suggest in what follows that rights-talk is vitally necessary, but also that it is insufficient as a resource for social movements. I also want to stress that in some ways strengthening the state is also vitally important—especially when it comes to the role of regulation—but that this too is inadequate. There is also the need to create numerous other sorts of *oppositional* institutions and strategies, some of which I will touch on. Finally, the discourse on rights examined above has neglected the issue of *scale,* and I will suggest in what follows that an understanding of scale is essential, both for academics seeking to analyze and suggest strategies and for activists engaged in strategizing, as they work toward the democratizing (and thus for the proper strengthening) of the state so as to effectively socially regulate capital and create useful and necessary rights for people.

■ What Rights Do; Or, the Production of Space for Democratic Ends

It is commonplace, but also inaccurate, to assert that "discourse" *produces* things—whether those things are bodies, social practices, or whole cultures. Yet this is not to say that discourses have no power. Quite the contrary, discourse helps set the context within which practices occur and are given meaning. Yet the power of words organized as discourses lies mainly only in this ability to *instruct*. Take the example of legal discourse. Laws and the discourse surrounding them can seemingly do all sorts of things. As laws, they can grant freedom or deal in oppression; they can order and regulate; or they can lead to mayhem. Yet in actuality, it is not at all the legal words that do this. Words alone do not prevent striking workers from engaging in secondary boycotts; words alone do not prevent gays from engaging in sex; words alone do not prevent women from attending military school or engaging in combat; words alone do not permit a corporation from taking subsidies to locate their plants in certain communities only to pull up stakes a few years later leaving a wake of destruction behind them. Rather it is *police power* that allows for these outcomes. At most, words can instruct and perhaps restrain that police power; they can help define other institutions of power that may provide a check on the police power of the state. In this sense, words can provide an invaluable tool for restraining power, for arraying it in this way and not that. That is precisely what "rights" do: They provide a set of instructions about the use of power.[2]

Tushnet (1984) argues that rights actually do very little because of their indeterminacy. "To say that rights are politically useful is to say that they *do* something, yet to say that they are indeterminate is to say that one cannot know whether a claim of right will do anything" (p. 1384). Yet Tushnet here ignores the way that a claim of right, *no matter how contested,* establishes a framework within which power operates. It matters less that power may breach this framework as often as honor it, because it is precisely there that the political utility of rights-talk does come to the fore. The argument (such as Tushnet would make) that such claims of right cannot be determined within the discourse of rights is correct. Adjudication of the abuse of power within the framework will always be a *political* action. It cannot be otherwise. But that does not thereby diminish the power of rights-talk, as Tushnet claims it does.

For example, how would Tushnet, with his example of the need for food and shelter, react to the widespread adoption of antihomeless laws around the country (Mitchell, 1997)? Surely, he would agree with Jeremy Waldron's (1991) startlingly obvious assertion in this context that "No one is free to perform an action unless there is somewhere he is free to perform it" (p. 296). No matter how appalling it might be to argue in favor of the right to sleep on streets (or urinate in alleys, sit on sidewalks, etc.), it is even more appalling, given the current ruthless rate at which homelessness is produced in society, to argue that homeless people should *not* have that right (Mitchell, 1997). That is, to the degree that we deny homeless people the right to sleep on sidewalks, we reinforce the "right" of the housed to never have to see the results of the society they are (at least partially) culpable in making. By denying the right to sleep, defecate, eat, or relax *somewhere,* Waldron (1991) concludes, contemporary antihomeless laws—predicated as they are on the rights of property—simply deny homeless people the right to *be* at all. In this instance, then, the denial that rights do anything (even if not autonomously) is genocidal. Likewise, absent the institutional power that rights-talk helps organize and constrain, it is hard to see how Rorty's call for compassion and moral persuasion will have any purchase against antihomeless laws that take as their basis the twin "commonsense" notions that property *must* be protected and that there is no reason why people should urinate and sleep in parks and on streets.

Or, more to the point, moral arguments create no institutions or institutional contexts that could protect that very moral argument. What happens if, in a few years, society does not produce homelessness quite so ruthlessly, and it is not quite such a bad problem? Or, for that matter, how does morality protect against the deprecations of a malicious (or simply selfish) majority? It should be clear, then, that rights-talk, among so many other processes, creates the context within which social worlds are created. Hence, while *Romer v. Evans* did not overturn *Bowers v. Hardwick,* it did, nonetheless, create a rhetorical and, more importantly, an institutional context within which struggle for the decriminalization of homosexuality may occur. As with *Brown v. Board of Education* of a generation before, *Romer v. Evans* was made possible by long, arduous political struggle of the type Tushnet advocates. But also like the Black Civil Rights movement, the gay rights movement now has a more sturdy peg on which to hang its claims and around which to organize its political activities.[3]

It is helpful in this regard to understand the institutionalization of rights (or more generally the establishment of laws) as a moment in the production of space—and I mean here material, physical space, not only the sort of metaphorical social space that has so intrigued scholars of the Left for the past decade or two. The "production of space" argument can be quite complex, but it is in its barest outlines that its utility for social movements seeking to regulate capital and create a just world may best be glimpsed. As numerous commentators have suggested (Foucault, 1980; Massey, 1984; Smith, 1990; Soja, 1989), space has been historically neglected as both a product and determinant of social action. Space, as Smith (1990) notes, has typically been "taken for granted, its meaning unproblematic" (p. 66). Space has been understood in its "absolute" or "abstract" sense—simply as a container, a blank surface, a void waiting to be filled. In the past two decades, however, that sense of space has begun to change as more and more analysts come to realize how absolutely central spatiality is to social life (e.g., Bourdieu, 1977; De Certeau, 1984; Giddens, 1981, 1984; Lefebvre, 1991). Most particularly, Lefebvre, Smith, and others have shown convincingly that absolute, abstract space is itself a social production, an outcome of social life. So too, of course, is relative space.

What is interesting is precisely the dialectic between these different types of produced spaces. Lefebvre (1991) traces abstract space from the "time that productive activity (labour) became no longer one with the process of reproduction that permeated social life," to when labor processes were divorced from the direct provision of daily necessities and social life (p. 49). Labor was "abstracted" from social life and in the process *abstract space* was produced. Lefebvre further argues that under capitalism, this abstract space has become predominant, transforming social life into a series of abstracted social relations. Because hegemonic, abstract space is the space of capitalism, it is the arrangement of space that makes capitalism possible. This is why Lefebvre (1991) argues that

> [I]t is struggle alone which prevents abstract space from taking over the whole planet and papering over all differences. Only the class struggle has the capacity to differentiate, to generate differences which are not

intrinsic to economic growth *qua* strategy, "logic" or "system"—that is to say differences which are neither induced by nor acceptable to that growth. (p. 55)[4]

The struggle for rights and for just laws is one aspect of this struggle to resist the hegemony of abstract space. But it is also a determinant in how space itself is produced. The rules for how capital moves across boundaries, for how firms develop in locations, for how public space is created, used, and transformed are all, in part, rules of law. Social action is structured through law, and social action creates abstract or differentiated spaces in proportion to the power each side in a struggle possesses. So social action—including oppositional work by social movements—always operates simultaneously to influence the production of law and the production of space. As Blomley (1994a) concludes, "Law is, as it were, produced in . . . spaces; those spaces in turn are partly constituted by legal norms" (p. 46). The struggle for rights—for example the right to sleep unmolested in a public park if you are homeless—becomes a clear tool, in the production of space, against powerful, abstracting forces (see Mitchell, 1992, 1995). But rights-talk is more than a tool; if successful, it provides institutional support for the maintenance of produced, differentiated space to be maintained against the continual forces of abstraction that seek to destroy it. Rights themselves, therefore, are part of the process of producing space. Lefebvre (1991) stakes out the end point of this argument: "A revolution that does not produce a new space has not realized its full potential; indeed it has failed in that it has not changed life itself, but has merely changed ideological superstructures, institutions or political apparatuses" (p. 54).

■ Scale and the Efficacy of Rights

In this sense, Representative Newt Gingrich is not exaggerating when he refers to the Republican program for Congress following the 1994 elections as a "revolution." For what is most clear—as the essays in this book show—is that new spaces *are* being actively produced. Most clearly, they are being produced by transforming the *scales* at which social, political, and economic life is conducted.[5] If one key front that social movements need to work on is *rights,* and another is the production of *space,* then a third is the production of *scale.* Mastery of scale is every bit as important as mastery of space. If the agents of capital have figured out the importance of space, then it is the agents of the state who have learned the significance of scale. Think of Franklin Roosevelt's strengthening of the nation-state at the expense of states, or the reverse trend instigated with Nixon and gaining so much momentum with Reagan, Bush, Clinton, and the Congresses of the 1990s. The importance of manipulating scale is not confined to actions of the state, however, for social movements within the large civil society (particularly those associated with the New Right) have learned to work at multiple scales.

At the outset of this chapter, I highlighted the coincidence of Rorty's essay and the Supreme Court's decision in *Romer v. Evans*. This coincidence clearly illustrates the centrality of scale issues in social movements, even if Rorty does not recognize the importance of these. Anti-gay rights activists have for several years been working simultaneously at multiple scales: on local ordinances, such as those in Cincinnati and numerous municipalities in Oregon; on statewide initiatives, such as Colorado's Amendment 2; and at the national level through the media (largely by instigating and perpetuating a set of "culture wars" that have transformed the ideological terms of debate). But thus far, gay rights—at least on the legal front—has remained a "local" issue. The statewide victory in Colorado prompted an immediate reaction in two ways. On the one hand, activists took the issue nation-wide, encouraging a boycott of the state and quickly organizing to halt similar movements elsewhere by learning from the mistakes of the anti-Amendment 2 forces in Colorado. On the other hand, legal challenges were initiated in both state and federal courts, seeking to overturn the amendment. The claim in federal court was simply that the federal equal protection clause of the Fourteenth Amendment "trumped" the state's right to limit political participation by some of its citizens. By jumping to the national scale, opponents of Amendment 2 were able to bring to bear a different set of legal resources than would have been possible only within the state.[6] But so too did the Right change scales as it sought to advance the anti-gay cause, expending most energy not at the state level, but in dozens of localities, forcing rear-guard actions by those seeking to defend gay rights in each and every one.[7] By moving down the spatial hierarchy and focusing action on the local state, the Right has been able to contest gay rights activists' move to the national scale.

But if this use of different scales for political ends is important, perhaps even more important has been the New Right's ability to create

State Restructuring and "Rights-Talk"   129

viable, interconnected *institutions* at all these scales, from the local parish to national (and transnational) networks of activists, ready to do battle on issues of abortion, gay rights, divorce, and school prayer. These initiatives are not limited to the cultural sphere, however; for affiliated groups, often well-funded by large corporations, have arisen to do battle against environmentalism, property-rights restrictions, tax laws, and campaigns to limit the mobility of capital. There are a number of lessons the Left can learn from the New Right—about organizing across scales, about the need to "capture" the state at all levels, about the importance of setting the terms of debate in the media—but there are also strong impediments to action.

For example, big capital is already mustering its forces in Congress to limit the ability of unions and other organizations from engaging in "corporate campaigns"—campaigns that seek to mobilize support for striking (or otherwise dissatisfied) workers by targeting all of a corporation's holdings and its corporate customers. Corporate campaigns seek out allies in the environmental movement to research a corporation's environmental record, linking workplace abuse with abuse of the environment in the vicinity of plants. They submit (or have allies submit) shareholder resolutions at general meetings, calling on corporations to reach fair settlements with their workers and their communities. They make common cause with consumer groups and support boycotts of products made by subsidiaries or fellow companies within the corporation they are campaigning against. And most significantly, they link their struggles in the United States to similar struggles overseas, publicizing environmental records in areas where corporations seek to locate new plants, for example.[8] In essence, union locals have been using corporate campaigns to "jump scales" (Smith, 1992)—to move their struggles beyond their seemingly isolated locality and to link it to struggles across the nation and across the globe.

The corporate response has been vehement, seeking to contain precisely that ability to jump scales—seeking, that is, to ensure that struggle, if allowed at all, remains only local and disconnected from similar struggles (perhaps against the same company) elsewhere. One tool to this end has been to file federal-level RICO suits against unions, alleging that they are engaged in a conspiracy to interrupt business. Another tool has been to work with Congress in hopes of simply outlawing corporate campaigns completely (Press, 1996, p. 32)—to entrap political activism at the scale of least efficacy (see Herod, this volume). Yet here too, the language of rights becomes vital. As the head of the

AFL-CIO's Industrial Union Department, Joe Uehlein remarked, "Corporate campaigns are based on the First Amendment and on the free flow of truthful information. . . . Sometimes the truth hurts" (quoted in Press, 1996, p. 32). This is not just empty rhetoric. To the degree that corporate campaigns can be defended as a right to expression at the national level, then local scale-jumping activism can be protected from corporate attempts to contain it.

Here we come to an intriguing argument about scale and the rhetoric of rights recently made by Nicholas Blomley (1994b). He argues that *where* rights claims are made is crucially important. Assessing the arguments of Tushnet, Robert Unger (1983), and other "rights skeptics" associated with the critical legal studies movement, Blomley agrees that rights-talk can backfire. Even so, he takes issue with the rather universal claims against rights made by legal scholars and argues instead that the effectiveness of rights claims depends largely on the situation in which they are made. As he puts it, "[w]hen making rights-based arguments . . . activists need to be careful in clarifying the location within which rights are put to work" (Blomley, 1994b, p. 413). What this suggests to Blomley is that *both* sides are correct: "For optimists, rights acquire meaning and progressive potency when deployed in *community* settings as mobilizers and political yardsticks; for the pessimists, it is the circulation of rights within the *juridical* domain that ensures that their meaning is counter-progressive" (Blomley, 1994b, p. 413). But this argument, by restricting the efficacy of activism around rights to the local (for the phrase "community *settings*" is crucial to Blomley's meaning), concedes too much. It gives away the larger scale arenas of struggle that may be quite crucial for establishing the context of rights in community settings to precisely those who refuse the progressive potential of rights, and it ignores the power of the judiciary (and other lawmaking and deciding bodies) as institutions that need to be won over if progressive causes are ever going to be able to have more than a local purchase.

■ Globalization, Devolution, and the Redefinition of Rights

What is at stake here is less the *where* of rights—because that is a battle that must be engaged on all fronts and at all scales—than the *definition* of rights and of what has rightful standing in society. I say *what* and not *who*, because that is precisely where the problem lies. It is easy to forget, and the popular media certainly make no effort to help us remember, that capital is entirely a creature of the state. It has no existence but that which is socially sanctioned through regulation or the lack of regulation. Capital—and thus companies and corporations—are not people. Although a person is certainly regulated and very much a creature of the state in many regards, it is nonetheless the case that a person is also something else simply by virtue of being human. This is not true of corporations. In the United States, corporations exist by dint of a state-issued charter that allows them to be treated as individuals for legal purposes. Article I, Section 10, of the U.S. Constitution provides that "No state shall . . . pass any Law impairing the Obligation of Contracts," and in *Dartmouth College v. Woodward* (1819), the Supreme Court decided that corporate charters were contracts between the state and investors. In essence this should suggest that once a charter is issued, the state has little or no authority to further regulate the activities of the corporation.[9] But by 1837 in *Charles River Bridge v. Warren Bridge,* the Court upheld the state's right to retain at the time a charter is issued a right to amend that charter at a later time. The majority's reasoning in this case centered around the need for the state to protect public interest. Although the "rights of private property must be sacredly guarded," Chief Justice Roger Taney wrote for the majority: "the object and end all of government is to promote the happiness and prosperity of the community" (*Charles River Bridge v. Warren Bridge,* 1837, p. 548). Even so, "[f]or a century after *Charles River Bridge,* the Supreme Court had little direct involvement with the law of corporations" (Wiecek, 1992, p. 199) with one important exception.

In 1886, writing a preface to a ruling concerning the right of localities to tax railroad corporations, Chief Justice Morrison Waite asserted that "we are all of the opinion" that the equal protection clause of the Fourteenth Amendment applies to corporations (*Santa Clara County v. Southern Pacific Railroad Co., 1886*). That is, corporations were to be treated under the law as "natural persons" with all the rights and privileges enjoyed by citizens. With this, as Wiecek (1992) summarizes, "The Court's substantive due process and freedom of contract decisions between 1890 and 1937 strengthened the hand of corporations in their dealings with employees, unions, consumers, and state legislatures" (p. 199). And matters have hardly changed since. Indeed, the rather bizarre ruling that corporations are "natural persons" has been used as a bulwark against all manner of regulations made in the public interest, from public access to the airwaves (because a corporation's "freedom of speech" would thereby be abridged if they were forced to air views with which they did not agree) to hindering the flight of capital from communities (because a corporation's "right to travel" would be hindered).[10]

What is at stake then, is precisely the legal definition of "person," and thus of citizenship. To the degree that the Supreme Court recognizes a corporation as a "person," it also imbues it with the same rights that people have. But as a total creature of the state, granting such rights to corporations is sheer nonsense. It misconstrues *what* is capable of possessing rights. But just as important, it makes the globalization of capital appear wholly natural, when it is nothing of the sort—when it is, in fact, just as much a creature of state decision making as is the creation of a corporation itself. And this is true at all scales: "free trade" is the creature of treaties; the ability of capital to abandon localities is the creature of a lack of regulation—and nothing more. Unless we confuse corporate capital with people, we cannot assume that capital or corporations have any rights whatsoever, except those specifically granted to them by the state.

Consider this in light of Kodras's remarks at the beginning of this section:

> The current drive toward global economic integration in the midst of a two-decade slowdown in productive expansion compels change in the operation of firms and governments placed around the world as each seeks to reposition itself within the intensifying competition. The result is a fundamental transformation in the global political economy, a rearrangement of the *geographic* relations between capital, the state, and civil society. Of the three, *capital* [is] the most mobile. (p. 89)

It clearly does not have to be. Its very mobility is allowed for by the state and civil society.[11] The right to mobility is *granted* to corporations, and by no stretch of philosophy can it at all be argued that, as with people, there might very well be "natural" rights that attach to them.

Kodras continues her analysis by arguing that nation-states are increasingly unable to control the global economy of which they are now a part—and local states even less so. But again, this does not necessarily have to be the case. There is nothing "natural" about a state's impotence. A sustained movement (as was begun in opposition to NAFTA) to reclaim corporations as creatures of the state, and thus as entities responsive to the public good, would go a long way toward finding ways to regain control, not only at the level of the national and local state, but also through global regulatory agencies. So finally, there is one further front on which the struggle for rights must be waged, and that is precisely at the global scale. If the restructuring of the state and the globalization of capital is ever to be properly socialized, then notions of global democratic citizenship (for people) will have to be forged. Capital has to be fought at its own scale, and the global institutions charged with "regulating it" (like the World Trade Organization) must be democratized and made accountable, not to capital and its bought politicians under the name of unfettering the economy, but to people under the name of *re*fettering capital.

■ Conclusion: It's Not Enough

Such arguments might advance goals that seem unattainable. Yet the aphorism "Politics is the art of the possible" should be reversed to "The possible is the art of politics." For possibilities to be invented and implemented, the Left needs to continue the hard work—against all manner of odds (corporate power, oligarchic control of the media, the continued success of authoritarian populism in organizing consent for repressive practices)—of creating institutions that can sustain progressive change. Lessons from the recent strikes and lockouts in Decatur, Illinois, present perhaps the clearest example of the type of the work that must be done, and of the difficulties attendant on sustaining innovative social movements for progressive change. In a city of 84,000 people, nearly one in four families had a member engaged in a strike or lockout in 1995. Some 1,800 workers were striking Caterpillar; another 1,250 were on strike against Bridgestone-Firestone; and 700 A. E. Staley workers were locked out in a bitter dispute about rotating 12-hour shifts and work rules. Workers saw themselves as residents of a "self-proclaimed War Zone of the new global economy" (Cooper, 1996, p. 21). Yet despite a remarkable radicalization of many local union members, innovative corporate campaigns against Staley clients (such as Pepsico and Miller Brewing Co.), support from left-leaning activists around the country, and even a bit of interest from the national media (which has proven adept at ignoring labor struggles), Decatur workers lost all three labor struggles, returning to work without contracts after bitter, long campaigns. Sold out by their internationals (which succumbed to sweetheart deals), those most radicalized by their struggles found it increasingly difficult to bring rank-and-file members along with them. One local leader put it this way:

> We won the national solidarity battle, but we lost the struggle on the shop floor. We're all corn-fed Americans, and at first we didn't know how to counter the company. We would be forced to escalate our strategy in order to hold it back. But every time we did, our people would get afraid. . . . We lost fighting for democracy. The more progressive we got, the more separated we got from the rank and file. (Cooper, 1996, pp. 23-24)

In the words of labor reporter Marc Cooper, "That it takes a drama of almost biblical proportions to radicalize a couple of dozen trade unionists tells us much about the state of class consciousness in America" (Cooper, 1996, p. 24). He points out that the activists he interviewed in Decatur "all grew up in union families in a quintessential union town," but that there still was no institutional (or cultural) support to draw on for sustenance in the long days of struggle:

> American workers not only have no candidates or parties that represent labor, but they have no media, no social memory, no culture that preserves or transmits traditions of class-based politics. The left rarely helps, given its often fixed obsession with identity politics, flunking workers on a cultural litmus test without ever dealing with them. (Cooper, 1996, p. 24)

One Decatur activist was more direct:

> I'd never heard of Joe Hill till someone showed me where he was executed. Someone else showed me where the Pinkertons shot down union workers. Gary and I were in Chicago one night and we end up at the Eugene Debs Dinner and there we are sitting at the gay rights table! (Cooper, 1996, p. 23)

Clearly, it is not as if institutions—such as unions—capable of channeling the energies of local activists, do not exist; but it is the case that they have little purchase on most workers' lives. That is due, in large degree, to the wholesale adoption of what used to be disdainfully called "business unionism"—the sort of unionism based on exclusion and privilege for a select few traded in exchange for toned-down activism, limited claims on the "rights" of the company, and the tacit agreement to separate labor issues from all the "cultural" issues that mark our lives. Into the void created by labor when it abandoned all hope of being a social movement has flowed a wide range of other institutions seeking to create for working people institutions of memory and meaning. Cooper concludes his report on Decatur by visiting the congregation of the Glad Tidings Assembly of God, which during the years of labor struggle in Decatur grew from a handful of Pro-life activists to more than 500 members. "Here one finds a cradle-to-grave operation, providing all the social services and community once offered to immigrant groups by labor organizations. One also finds a lot of blue-collar workers here, many of them active union members" (Cooper 1996, p. 25). According to one member, Glad Tidings has become successful because "Pastor has taught us the value of team-work. Individuals are powerless. He has taught us that in union there is strength" (Cooper, 1996, p. 25).

The point is simple: progressive change—including the ability to redirect the state and to create in it a force for the protection of the rights of ordinary people instead of the putative rights of capital—begins first as a process of institution building. It is just as clear too that neither "cultural" nor "economic" struggle can be divorced from each other. What Glad Tidings and other churches do so effectively is provide an institutional framework for giving meaning to people's lives—meaning that helps them make sense of the economic, political, and social maelstrom of which they are a part, but also meaning that allows them, in one fashion or another, to *act*. There is no reason why the economic and cultural Left cannot adopt similar strategies for clearly progressive causes. However, if such institution building is going to be able to "jump scales" effectively—linking the struggles in Decatur, for example, with similar struggles against the prerogatives of capital in Germany or Nigeria[12]—then it must be based necessarily on an evolving language of rights. For it is only by recognizing the universality of "rights" that the particularities of place and people with which the Left has been for so long concerned can be protected against the rapacious forces of capital, forces made all the more potent by the disempowerment of the state (and of people organized through the institutions of the state) in the name of "devolution." Through a universal language of rights, backed by a complex network of progressive institutions working within and across scales (and producing new scales of action of their own), the abstract space of capital can be strongly contested in the name of difference, in the name of creating what is right. And if this is a differential space that is to take hold, if it is to truly challenge the homogenizing prerogative of capital, then it must necessarily configure that difference in noneconomic as well as economic terms. That is why using the state to protect "cultural" rights—such as the right to be gay—is as essential to progressive economic and political change as is activism geared toward stripping capital of its putative rights to determine our worlds. And make no mistake, as the debates over welfare and the criminalization of the "reserve army of labor" make so clear, the sort of "authoritarian populism" that excuses gay-bashing is hardly unfriendly to the totalizing ambitions of capital (even if Patrick Buchanan represents an interesting exception).

In the struggle to retain at least some control over their economic and social lives, the workers in Decatur tried the "appeal to fraternity" that Rorty recommends as an alternative to the appeal for rights. And that fraternity has been an important, if contradictory, aspect of their struggles to fashion appropriate institutions to stand up to the threats that confront them. But in the face of capital that has been granted the "right" to hire permanent scabs, to put workers on debilitating 12-hour rotating shifts (see In the quest for efficiency, 1996), or to simply close shop and put thousands out of work, such fraternity is clearly insufficient. Institution building is one thing; institutional *protection,* quite another. Without a language of rights to defend that institutional protection, to define its power, fraternity will prove no match against the power of capital and its cultural allies (who have it exactly backwards by wanting a laissez-faire state for money but not for people) to redefine the state in their own terms.

NOTES

1. Blomley (1994b) forcefully makes just this claim. He further argues (following Williams, 1991) that the Left critique of rights is quite condescending when the experience of African-Americans, for example, whose everyday experience is one of the denial of their rights rather than the protection of them, is taken into account.

2. I have repeated here an essentially negative argument about rights—that rights are for protection *from.* But the argument is the same for positive rights—police power is essential in establishing contours for the provision of rights *to* things. The right to a place to sleep, for example, requires police and regulatory power just as much as the right to protection *from,* for example, an invasion of privacy.

3. And like *Brown v. Board of Education,* whatever rights may be protected in *Romer v. Evans* are not secure as social conditions change. For Tushnet, this is an argument against rights-talk. For activists, it is an argument for vigilance over rights won.

4. In a move that many contemporary analysts would find problematic, Lefebvre includes "political action of minorities" under the rubric of class struggle. He is right to do so in some senses, since struggles for liberation by African-Americans or gays or

post-colonial peoples cannot be divorced from the political-economic system in which we live.

5. It is as easy to take scale for granted as it is to take space for granted. But like space, scale is actively produced rather than given. The movement for devolution and for the unfettering of capital at the global level makes that painfully clear.

6. Indeed, the majority ruling in *Romer v. Evans* was made on quite different grounds than those advanced by the State Supreme Court.

7. The battle has gone national again, with the impending decision in Hawaii on gay marriages, the controversy over gays serving in the military, and the decision in *Romer v. Evans*.

8. For a good description of these campaigns, see Press (1996) and Herod's chapter in this volume. For an argument concerning the ways in which capital constructs and uses scale to contain labor, see Mitchell (in press).

9. It should be noted that until very recently, the Supreme Court has to a large extent deferred to states on matters concerning the nature of corporate charters, but that since the passage of the Securities Exchange Act of 1934, the federal judiciary has considerably expanded its reach.

10. See Blomley (1994b) for an intriguing analysis of ways in which "right to travel" provisions can be used to make capital *stay put*.

11. To the argument that regulating corporations in the United States—and hindering their movement—will simply make them flee overseas, one only has to consider the size of the U.S. market. A double-edged regulation, which required firms selling in the United States to abide by the standards of the United States would go a long way toward remedying that problem.

12. I have chosen these two examples—Germany, with its racist near-war being waged against "guest workers," and Nigeria, where the ability of Shell and other oil companies to operate depends on the perpetuation of ethnic divisiveness—to remind us that economic and cultural struggle are interdependent, so it is invidious to make too much of distinctions between the economic left and the cultural left, or between economic conservatives and cultural conservatives. It is not as if they exist in separate worlds defined by separate social processes.

REFERENCES

Blomley, N. (1994a). *Law, space, and the geographies of power.* New York: Guilford.
Blomley, N. (1994b). Mobility, empowerment and the rights revolution. *Political Geography, 13,* 407-422.
Bourdieu, P. (1977). *Outline of a theory of practice.* Cambridge, UK: Cambridge University Press.
Bowers v. Hardurck 478 U S. 186 (1986)
Charles River Bridge v. Warren 36 U.S. 420 (1837)
Cooper, M. (1996, April 8). Harley-riding, picket-walking socialism haunts Decatur. *Nation,* 21-25.
Dartmouth College v. Woodward 17 U.S. 518 (1819)
De Certeau, M. (1984). *The practice of everyday life.* Berkeley: University of California Press.

Foucault, M. (1980). Questions on geography. In C. Gordon (Ed.), *Power/knowledge: Selected interviews and other writings, 1972-1977* (pp. 63-77). New York: Pantheon.
Giddens, A. (1981). *A contemporary critique of historical materialism.* Berkeley: University of California Press.
Giddens, A. (1984). *The constitution of society: Outline of the theory of structuration.* Berkeley: University of California Press.
Hall, S. (1988). *The hard road to renewal: Thatcherism and the crisis of the Left.* London: Verso.
In the quest for efficiency, factories scrap 5-day week. (1996, June 4). *New York Times,* pp. A1, C19.
Lefebvre, H. (1991). *The production of space.* Oxford, UK: Basil Blackwell.
Massey, D. (1984). *Spatial divisions of labor: Social structures and the geography of production.* London: Macmillan.
Meszaros, I. (1995). *After capital: Towards a theory of transition.* New York: Monthly Review Press.
Mitchell, D. (1992). Iconography and locational conflict from the underside: Free speech, people's park, and the politics of homelessness in Berkeley, California. *Political Geography, 11,* 152-169.
Mitchell, D. (1995). The end of public space? People's park, definitions of the public, and democracy. *Annals of the Association of American Geographers, 85,* 108-133.
Mitchell, D. (1997). The annihilation of space by law: The roots and implications of antihomeless laws in the United States. *Antipode, 29,* 306-338.
Mitchell, D. (in press). The scales of justice: Localist ideology, large-scale production, and agricultural labor's geography of resistance. In A. Herod (Ed.), *Organizing the landscape: Geographical perspectives on trade unions.* Minneapolis: University of Minnesota Press.
Press, E. (1996, April 8). Union do's: "Smart solidarity." *Nation,* 29-32.
Romer v. Evans 116 S. Ct. 1620 (1996)
Rorty, R. (1996, June). What's wrong with "rights"? *Harper's, 202,* 15-18.
Santa Clara County v. Southern Pacific Railroad Co. 118 U.S. 344 (1886)
Smith, N. (1990). *Uneven development* (2nd ed.). Oxford, UK: Basil Blackwell.
Smith, N. (1992). Contours of a spatialized politics: Homeless vehicles and the production of geographical scale. *Social Text, 33,* 55-81.
Soja, E. (1989). *Postmodern geographies: The reassertion of space in critical social theory.* London: Verso.
Tushnet, M. (1984). An essay on rights. *Texas Law Review, 62,* 1363-1403.
Unger, R. (1983). The critical legal studies movement. *Harvard Law Review, 96*(3), 320-432.
Waldron, J. (1991). Homelessness and the issue of freedom. *UCLA Law Review, 39,* 295-324.
Wiecek, W. (1992). Corporations. In K. Hall (Ed.), *The Oxford companion to the Supreme Court of the United States* (pp. 198-199). New York: Oxford University Press.
Williams, P. (1991). *The alchemy of race and rights.* Cambridge, MA: Harvard University Press.

Part III

Implications of State Change

8

FAIR or Foul? Remaking Agricultural Policy for the 21st Century

BRIAN PAGE

Early in 1995, the 104th Congress set out to work on a new farm bill. In July, House Agriculture Committee chair Pat Roberts—following the lead of his counterpart in the Senate, Richard Lugar—began to promote a new concept in farm legislation, the "Freedom to Farm" Act. This proposed legislation departed from traditional farm policy by eliminating most federal price and income supports of farm production over a 7-year period. It represented a radical break from past governance of U.S. agriculture. In November 1995, both the House and the Senate passed the Freedom to Farm legislation and sent it to President Clinton as part of the 7-year balanced budget bill. The Republican budget bill was vetoed by the President. Further action on farm legislation waited until February 1996, when the Senate passed a somewhat modified version of Freedom to Farm before recessing. Back from recess in early March, the House followed suit; and later that month, a Senate-House Conference Committee ironed out differences between the two bills and agreed to bring the compromise Farm Bill to a vote. Both the House and the Senate approved the new legislation and on April 4 President Clinton signed into law the Federal Agricultural Improvement and Reform (FAIR) Act of 1996. Despite its route through a complicated political process characterized by trade-offs, vote-buying, and all the rest, the FAIR Act looks remarkably like the original Freedom to Farm concept proposed by Pat Roberts in the summer of 1995.

It is important to remember that the restructuring of agricultural policy was not a cornerstone of the much-publicized Republican *Contract with*

America. Nonetheless, Roberts' proposal was enthusiastically supported by the Republican leadership because it fit perfectly with their broader goal to redefine what constitutes the appropriate function of the federal government. Indeed, the very language of the initial legislation cast federal involvement in agriculture as an enormous liability to the success and prosperity of American farmers. So too, the FAIR Act echoes the strong free-trade, antisubsidy position taken by U.S. representatives to the Uruguay round of the General Agreement on Tariff and Trade (GATT), a stance that was recently codified in the Uruguay Round Agreement (URA). The new farm legislation is now listed among the significant accomplishments of the 104th Congress.

This chapter provides an overview and analysis of the FAIR Act of 1996. Farm policy is certainly distant from the everyday lives of most (urban) Americans, and it may seem quite peripheral to the main arenas of discourse concerning state restructuring. Nevertheless, the creation of a new Farm Bill is representative of some of the key processes through which relationships between the state, capital, and civil society are being rearranged. In the case of agriculture, restructuring has occurred mainly through efforts to dismantle the existing regulatory system. Yet, while some components of federal farm policy have been eliminated, others have been retained and expanded. This pattern of partial and selective dismantling reflects political struggles over what constitutes the appropriate function of the federal government in agriculture, and, by extension, whose interests should be served by farm policy. Despite the fact that the FAIR Act employs the rhetorical sleight-of-hand that has become the stock-in-trade of American politics, this process has clear winners and losers. Indeed, I argue that the FAIR Act is anything but fair, because it "improves" and "reforms" for the benefit of the few. The new agricultural policy framework represents a clear victory for corporate, globally oriented agribusiness at the expense of the social and economic health of rural America. Moreover, as one of the first pieces of legislation signed into law after the government shutdowns of late 1995, farm policy can be used as a bellwether of political concessions and alignments that may appear in future legislation if Congress continues to undertake a broad restructuring of U.S. social and economic policy.

The chapter begins with a discussion of the new Farm Bill in relation to previous farm policy and then turns to several aspects of the FAIR Act that reveal the perversity of its packaging—I refer to these as the FAIR Act Ironies.

■ Restructuring Farm Policy: The FAIR Act of 1996

The backbone of the FAIR Act is the Agricultural Market Transition Program, which lays out a framework that replaces traditional subsidies with "Production Flexibility Contracts."[1] This new system uncouples farm subsidies from the market for most major program crops, including feed grains, wheat, cotton, rice, and oilseeds. Farmers of these crops will no longer be paid to enroll in the Acreage Reserve Program (ARP) following years of surplus. The ARP was originally intended to keep land out of production for certain crops in order to buoy market prices. Nor will farmers receive deficiency payments when market prices plunge below government-established target prices.[2] Freed from the planting restrictions of these commodity programs, farmers will be able to plant any crop except fruits and vegetables on 85% of their "contract acreage"; they can plant without any restriction on the remaining 15%.

In addition, farmers will receive a guaranteed severance payment (amounting to $36 billion) to ease this market transition. These payments are based on subsidies received on program commodities over the past 3 years and will be made to farmers over a 7-year phase-out period regardless of market trends. In order to receive payments or loans on program commodities, the Production Flexibility Contract requires a farmer to comply with existing conservation plans, wetland provisions, and flexibility provisions, and to keep the land in agricultural uses.

While the FAIR Act eliminates both the ARP and target prices, one key price-support mechanism remains in place—nonrecourse loans. The nonrecourse loan program was initiated in the 1930s (Cochrane, 1979). It provides operating capital to farmers of wheat, feed grains, cotton, peanuts, tobacco, rice, and oilseeds through the Commodity Credit Corporation (CCC). To obtain a loan, farmers put up a portion of their crop as collateral. Within a specified period of time, the farmer has the option to either repay the loan and reacquire the commodity or forfeit the commodity to the CCC. The government has *no recourse* but to accept the commodity as payment. The decision to reacquire or forfeit, of course, is based on the relationship between the loan rate and the market price for the commodity in question. When the market price is below the loan rate, the farmer can forfeit the commodity and come out ahead—he or she achieves a "marketing loan gain." At the same time, the acquisition and storage of the commodity by the federal government blunts overproduction and thus keeps prices from falling further.

The FAIR Act modifies the loan program with the specific intent to minimize the potential for forfeiture to the CCC and thereby to reduce government stocks of farm commodities, lower storage costs, and encourage farmers to sell on the market. To do this, the new legislation sets loan rates at 85% of the preceding 5-year average price (dropping the high and low price) and sets maximum loan rates on both wheat and corn. In addition, should government supplies of wheat and corn accumulate, there are provisions to adjust the loan rate downward an additional 10%.

The FAIR Act also lays out significant changes for the dairy program. Price support for dairy products will be gradually phased out over a 4-year period and eliminated entirely thereafter. The price support program has traditionally been carried out through CCC purchases of butter, nonfat dry milk, and cheese that enable processors to pay dairy farmers the support price for milk. Another price support mechanism, milk marketing orders, remains in place but with some reforms. The new Farm Bill requires that the number of milk marketing orders be reduced from the current 33 to no more than 14 over the next 3 years. Federal marketing orders authorize localized agricultural producers to promote "orderly" marketing in their areas by controlling supply and quality and by pooling capital for promotion and research. Consolidation of the marketing orders is the first step toward breaking apart the regional regulation of dairy pricing that has acted to insulate producers from distant competitors.[3]

Although most crops will move rapidly along the path to market transition, a few will not. In particular, the federal role remained intact in peanut production (albeit with modifications) and was virtually unchanged in sugar production. The peanut program reserves most of the domestic market for U.S. farmers who are issued production quotas by the government. These quotas are also geographically "fixed" by law; in the major producing states, they can be sold, leased, and transferred only within a county. The federal government also guarantees peanut producers a minimum support price through a nonrecourse loan program. The FAIR Act reduced the loan rate for peanuts by 10%, eliminated the minimum national quota so that supported production levels could be adjusted downward to more accurately reflect domestic consumption, and allowed the transfer of quotas (up to 40% of county totals) across county lines within states. In addition, the marketing assessment for peanut producers (a per-unit fee assessed in order to share program costs with the government) was increased by a third. There were no provisions for phasing out the program.

The sugar program was even less affected. The key feature of the sugar program—the restriction of imported sugar via the tariff-rate quota[4] (TRQ)—remains in place. Import restrictions act to support U.S. prices at levels designed to minimize forfeiture of CCC loans—that is, when the TRQ is at 1.5 million tons (the minimum level set by the URA of the GATT), loans to U.S. producers are recourse loans, and the government does not have to acquire and store the crop; but when the TRQ is raised above that level (meaning that more imported sugar enters the domestic market at low tariff rates), then U.S. producers have access to nonrecourse loans. Unlike other commodities, the FAIR Act did not reduce the loan rate for sugar producers. However, both cane and beet sugar producers must pay a very small penalty on amounts forfeited to the CCC. The marketing assessment for sugar producers also went up, but this increase was less than for peanut producers.

The original House Freedom to Farm concept dealt with market transition alone and left aside the many other aspects of comprehensive Farm legislation such as trade, conservation, nutrition, credit, and research. However, splitting the Farm Bill in this way was not acceptable in the Senate, and thus the FAIR Act retained the broader framework of traditional farm legislation. Not all aspects of farm policy were reassessed, but significant changes were made across a range of programs.

In particular, the FAIR Act combines the dismantling of commodity programs with a renewed focus on agricultural market development. Agricultural trade provisions have been a central part of farm legislation (and U.S. foreign policy) since the Agricultural Trade Development Act of 1954, known as Public Law 480. P.L. 480 sought to expand foreign markets and promote economic development by subsidizing foreign purchases (via low interest rate loans), providing "food for development" grants, and making food donations in cases of hunger emergencies. P.L. 480 and most other agricultural trade programs were reauthorized, and many new export-enhancing provisions were added. The key change was an increased emphasis on supporting the export of high-value and value-added products in addition to staple products. The USDA was also given the charge to coordinate and prioritize among export programs via the establishment of an agricultural export promotion strategy. In some cases, particularly dairy production, the expansion of export programs represents an effort to offset the shock of market transition.

The FAIR Act also retained and expanded the environmental programs of previous farm legislation—despite the fact that many environmental advocates initially viewed the Freedom to Farm concept as a threat to conservation practices in agriculture. The Conservation Reserve Program (CRP) was maintained at the current level of 36.4 million acres. CRP participants are allowed to terminate contracts prior to their expiration (10 years), but the funds saved through termination can be used to enroll new lands. In addition to taking environmentally sensitive lands out of production, the CRP also acts as a price-support mechanism because it takes acreage out of production. However, conservation compliance was weakened by provisions that allow producers to change the conservation practices set out in their conservation plans and give the Secretary of Agriculture greater discretionary power in exercising enforcement and establishing penalty levels.

Likewise, the Wetlands Reserve Program was retained, but was modified to (a) promote temporary easements that will allow farmers to drain and till wetlands, and (b) expand cost-sharing arrangements for restoration. At the same time, reforms to the enforcement of wetlands land use by the Natural Resources Conservation Service make it easier for producers to receive exemptions from the prohibition to drain previously undrained lands. As with conservation compliance, the Secretary of Agriculture was given discretionary power to establish penalty levels.

The FAIR Act also creates an Environmental Quality Incentive Program (EQIP) designed to assist crop and livestock producers in improving environmental management. This program is targeted at livestock producers and the growing problem with animal waste disposal and treatment. Assistance to individual operations is capped at $10,000 a year for a maximum of 5 years. Large operators—as defined by the Secretary of Agriculture—are ineligible.[5] Other environmental provisions of the FAIR Act include a $35 million Farmland Protection Program designed to preserve farmland from commercial development via the purchase of conservation easements and a $200 million allotment to purchase and restore land in the Florida Everglades. An additional $100 million in federal support for this program is contingent on the sale or swap of other federally held lands in Florida.

Changes were also made to the agricultural credit program. This program has two main components. The first is a set of farm loan programs administered by the USDA's Farm Service Agency (FSA), formerly the Farmers Home Administration. These programs provide real estate, operating, and emergency loans to individuals whose primary business is farming. Existing farm lending programs were reauthorized with some changes. The authority to make loans for nonagricultural purposes was repealed, and new restrictions were placed on emergency loans. Moreover, borrowers with delinquent accounts now face tighter loan restructuring rules, forfeited property can be sold more rapidly, and the USDA can use collection agencies to recover delinquent loans. The second component is the Farm Credit System (FCS), a combination of cooperatively owned financial institutions that finance farm and farm-related mortgages and operating loans. No changes were made to the FCS, but the FAIR Act directs the USDA to undertake a study of the adequacy of FCS services.

The FAIR Act strengthened the rural development provisions of previous farm bills. Traditionally, these funds were earmarked for infrastructural improvements in water, sewer, and telecommunication systems and for economic development activities. The FAIR Act maintained this orientation, but in addition created a $300 million "Fund for Rural America" to be used to augment existing resources.

The only major components of farm policy left untouched were nutrition assistance and research. The Food Stamp Program was reauthorized for 2 years while Congress works on comprehensive welfare reform legislation. The commodity distribution programs (including the Soup Kitchen and Food Bank programs) and the Temporary Emergency Food Assistance program were also reauthorized. Similarly, the Agricultural Research, Extension, and Education programs were reauthorized for 2 years while the Secretary of Agriculture undertakes the development of a system to evaluate these programs. It is likely that Congress will press for the partial privatization of these latter programs.

■ The New Farm Bill: FAIR for Whom?

The overall cost of the 7-year FAIR Act is $47 billion, of which $36 billion will be paid out as guaranteed payments to farmers. The cost of the 5-year 1990 Farm Bill was $53 billion. The FAIR Act was promoted as a free-market breakthrough that would reduce the role of government in economic life, cut federal spending, and move the nation closer to a balanced budget. It was also promoted as a commonsense approach to agriculture that would be more fair to farmers and consumers alike. In reality, however, the FAIR Act remakes agricultural policy in ways that belie this packaging.

FAIR Act Irony #1: A Massive and Temporary Buyout

The most glaring irony is that the FAIR Act will in all likelihood cost taxpayers much more than the traditional system of government planting restrictions and subsidies. There are two reasons for this. First, by dismantling most commodity programs now, farmers can keep $1.7 billion in subsidies advanced last year before commodity prices rose above target price levels. Under previous legislation, farmers would have to return this money to the government. Second, farmers will receive $5.6 billion in guaranteed transition payments this year while also reaping the benefits of higher commodity prices; farmers don't lose anything in the near term—in fact they gain an enormous windfall. Moreover, farmers will most likely continue to achieve unusually high incomes due to the combination of high prices with high "transition payments" for the next 2 or 3 years due to expected export growth that will keep prices high. In the old system, the costs of planting restrictions (acreage set-asides) and subsidies (target prices) would most likely be quite low under this set of economic circumstances. In February 1996, the USDA projected the cost of extending the current system of farm subsidies over the next 7 years at just $12 billion (USDA, 1996a).

In fact, the fortuitous timing of commodity price cycles was the key to the success of this Farm Bill. Pat Roberts is a quite recent convert to the free-market approach to agriculture. As recently as February 1995, he responded to Richard Lugar's proposed cutbacks in crop subsidies by calling the Indiana Senator "the Lizzie Borden of Agriculture" (Ingersoll, 1996). What accounts for Roberts' rather abrupt change in approach to farm policy is the fact that decoupling allowed farmers to retain their over-payments from 1995 and take an additional gulp from the government until 1996, 1997, and maybe even 1998. In reality, Roberts was not motivated by cutting government spending as advertised, but by acquiring as much government money for his constituents as possible through the exercise of classic "pork-barrel" politics.

Now there is risk in this type of short-term strategy, but Roberts is protected by his own lack of ideological commitment—he can switch back to a pro-interventionist stance as easily as he adopted the free-market stance. How? On the one hand, he allowed the 1949 Agricultural Adjustment Act to be retained as the permanent farm legislation. When the FAIR Act expires in 2002, Congress—through inaction—can simply revert to a traditional system of subsidies if farmers are in bad financial condition due to price downturns. On the other hand, Roberts held out the possibility of abandoning the FAIR Act even sooner. On April 1, this quote from Roberts appeared in the *Wall Street Journal*: "We hope it is the golden age of agriculture. If we go to hell in a handbasket, we'll have to say time out, get a budget waiver, and fix it" (Ingersoll, 1996). It would appear that the new Farm Bill is flexible in more ways than one! Added to this is the way that transition payments are structured over the 7-year Farm Bill. Payments float along at roughly the same level for 5 years and then drop like a rock. In this way, the benefits to farmers are maximized, the pain of market adjustment is deferred, and lip service can be paid to the goal of a balanced budget in 2003.

The question of whether or not these farm subsidies are actually paid to farmers as planned (let alone the question of whether or not they disappear in 7 years) remains, however. Seven years is a long time in politics, and a different Congress may well decide to approach farm policy in an entirely different way. In fact, the basic provisions of the FAIR Act are already under attack. In late May 1996, the House appropriations subcommittee attempted to cut $98 million in Production Flexibility Contract payments to farmers in fiscal year 1997. Agriculture committee members appealed to the Republican leadership and had the funds restored, but this indicates a very tenuous political commitment to the FAIR Act (Jones, 1996b).

FAIR Act Irony #2: Large-Farm Bias Is Locked In

Although the FAIR Act phases out farm subsidies, it does nothing to change a pattern of farm subsidy distribution that favors large farms. Large farms have traditionally garnered the lion's share of subsidies because farm program benefits are proportional to output—the larger the output, the larger the payment. By the late 1960s, the problems with this system became evident: In most commodity programs, the largest 20% of the farms—many owned by huge diversified corporations—received over half of the payments (Strange, 1986). In response, Congress imposed limits on the amount of money that could be paid out to individual farmers. Over the past 25 years, this limit has changed several times, and in 1995 it stood at $50,000. Large farms, however, have been able to effectively circumvent the payment limit through the use of the "three entity rule." This rule enables a farmer to split his or her farm into three separate operating entities as long as the farmer's ownership stake in the second and third entities does not exceed 50%. The farmer can then claim payments for each operating entity.

The FAIR Act made few changes in the allocation of farm subsidies. The current arrangements were locked in for 7 years, and no effort was made to redress the imbalance in the distribution of program benefits (targeting payments on the basis of need rather than production levels, for example). Congress reduced the payment limit on production flexibility contracts by 20% (down to $40,000 per person) and maintained a limit of $75,000 on marketing loan gains. The new legislation retained the three entity rule, the total allowable contract payments on three separate operating units was reduced from $100,000 to $80,000, and the total combined in marketing loan gains was set at $150,000. Current estimates are that just 2% of farm subsidy recipients (28,000 operations) will receive 22% of the total payments over 7 years. This is an average of $281,000 over 7 years—more than $40,000 a year in guaranteed payments with no requirement that these operations are still farming and no means test to determine need. This top 2% will be eligible for 11 times more in subsidies than average recipients (Environmental Working Group, 1996). Moreover, because there are no requirements that payment recipients be active farmers or farm managers, much of this money could potentially go to absentee landowners and nonfarmers living in urban areas.

FAIR Act Irony #3: The Uneven Geography of New Farm Policy

Making farm policy is a complicated process in which an array of geographically based political interests collide, negotiate, and ultimately come together in uneasy alliances. Of course, this involves party politics; but party distinctions often play a secondary role to conflicts between traditional commodity groups, between environmental groups and agri-business, between rural and urban interests, and between the House and Senate. The process of constructing farm policy is often one of geographic deal making.

What, then, are the geographic consequences of the FAIR Act? Does it live up to its name and achieve some sort of regional balance? Commodity-based regionalism almost crippled Roberts' farm legislation in the House. In fact, opposition to decoupling was most fierce from southern Republicans representing cotton and rice interests. Ultimately, Roberts had to move the bill out of the Agriculture Committee to the

FAIR or Foul? 151

Budget Committee in order to get it to the House floor for a vote. Nevertheless, the cotton and rice interests did manage to obtain higher price-support rates for their commodities and head off any attempt to take away the three entity rule. Another southern coalition fought off an amendment to the Farm Bill that would have cut back the peanut program, whereas the cane and beet sugar industries united to keep the sugar program virtually untouched. In the latter two cases, many of the representatives that authored amendments aimed at cutting the peanut and sugar programs were eventually pressured by the Republican leadership to switch sides and vote against their own amendments (Ingersoll, 1996).

The movement of the new farm bill through the Senate was also filled with regional conflicts and trade-offs. Senate Democrats from the largely nonfarm states of the Northeast were willing to support the market transition concept only if environmental programs were maintained and extended (Cushman, 1996). In the House-Senate conference, the key battle was over dairy policy, the Northeast's key agricultural commodity. Here, a compromise was reached that phases out dairy price supports but gives New England producers the opportunity to enter into a Dairy Compact if prices plunge too low.

Ultimately, the regional balance sheet favors the South. This is true, in part, because the FAIR Act bases its farm severance payments (Production Flexibility Contracts) on actual payments to farms from 1991 to 1995. Of course, the amount of money allotted to any given commodity program varies quite a bit from year to year according to production levels, market conditions, and so on. The past 5 years just happened to be a period of time in which wheat, but especially cotton, received a larger share of payments. Thus, new farm policy somewhat arbitrarily locks in a particular pattern of spending among commodity programs that favors one the of South's chief crops. At the same time, support programs for two other important southern crops, peanuts and sugar, were left largely intact.[6] These victories for southern commodities reflect a continuation of the South's historic lead role in the making of farm policy (Cochrane, 1979). In addition, several of the new environmental programs had a distinctly Southern orientation, including the efforts to restore Florida's everglades and to deal with North Carolina's growing livestock waste crisis.

The battle over the peanut and sugar programs involved a geographic rift between rural and urban representatives (with strong North-South overtones). In general, urban representatives pushed hard for the elimination of farm subsidies, despite the fact that a sizable chunk of commodity program payments (nearly $400 million) are made to absentee landlords residing in cities and to companies headquartered in major urban areas (Environmental Working Group, 1996). In particular, House representatives from urban districts took aim at the peanut program because it artificially raises peanut prices and thereby costs consumers an estimated $500 million extra per year. The same argument was marshaled against the sugar program, but both efforts failed in the face of southern Democrat-Republican unity based on rural interests.

The West also fared well in the new legislation. Again, this is due in part to commodity orientation, because both cotton and sugar beets are important western crops. Just as important, western fruit and vegetable interests headed off the emergence of some potential competition by making sure that key planting restrictions were maintained (to receive contract payments, farmers can plant fruits and vegetables on only 15% of their acreage). In this way, southern farmers in particular are prevented from switching to fruit and vegetable production. The West, with an agricultural sector dominated by corporate agribusiness, also benefited from the retention of the three entity rule.

Of the significant agricultural regions, the Midwest is likely to be the most adversely affected by new farm policy. Its major crop, corn, will receive a smaller share of the overall farm severance. Its dairy industry will be completely unplugged from government support. Its producers—which are smaller in scale and often family-based—will garner less of the farm severance and thus bear a disproportionately larger share of the burden of commodity program cuts. Most importantly, these producers will be much more vulnerable to the forces of market transition. This issue is taken up in the next section.

FAIR Act Irony #4: Cleaving Farm Policy in Two

For the operators of smaller farms, being released from the "heavy hand" of government regulation may ultimately hurt more than it helps. During the next few years, farmers will undoubtedly expand production in response to high prices and the lifting of restrictions. "Fencerow to fencerow" production in the next few years is a near certainty, and this will undoubtedly lead to overproduction and the depression of market prices. Proponents of the FAIR Act argue that strong and growing exports will counteract this trend to a significant degree. It is certainly true that international market conditions will now have a much greater impact on domestic prices than in the past. Yet most export projections are based on the assumption that exports to Asia will continue to increase (e.g., Food and Agriculture Policy Research Institute, 1996). This strategy of putting most of the eggs of the U.S. farm economy into the basket of Asian consumption growth seems more than a bit risky—particularly with respect to China. Added to this is the fact that it won't just be U.S. farmers attempting to tap foreign markets—the (partially) deregulated agricultural economies of Canada and Europe will be gearing up as well.

When (not if) a significant downturn in prices occurs, it will be small and mid-sized farm operations that are hurt—the traditional family farms that still occupy a central place in the pantheon of U.S. farming, particularly in the Midwest. Of course, larger producers will be affected, but these farm operators have several advantages that will allow them to absorb a loss in income, including better access to commercial capital, heightened efficiency through industrialization, and the cushion of large severance payments from the federal government. As a group, small and mid-sized farms will be much less able to handle a significant decrease in farm commodity prices. Most aspects of the former safety net for farmers have been eliminated. The one that remains—the non-recourse loan program—was altered in a way that gives farmers a price for their crops that will likely not cover their costs of production. Moreover, many of these farms lack the professional marketing and risk management skills that will be increasingly necessary for survival and success in the new market-driven environment of wide price fluctuations. In particular, the lower profit family farms will have trouble keeping their debt-to-asset ratios low in this context; and as this ratio rises, they will have more and more difficulty acquiring the capital necessary for continued operation.

Added to this is the fact that the credit provisions of the FAIR Act made important changes with respect to loan making, loan servicing, and borrowers' rights. FSA loans can no longer be used to refinance preexisting debts such as real estate mortgages or operating debts; nor can they be used to pay balloon payments on land purchase contracts. Delinquent borrowers can no longer receive an FSA loan, and they can buy out their indebtedness only at significantly higher levels. Any borrower who has received one farm debt forgiveness (consolidated, rescheduled, reamortized, or deferred) is no longer eligible for any direct or guaranteed loan. A borrower must now be current on all FSA loans in order to receive an annual operating loan. The program that allowed farmers to lease back their former properties from the USDA was eliminated; and now, when the USDA disposes of properties acquired through foreclosure, the previous borrower-owners are no longer given priority in the transaction. Finally, the ability of farmers to appeal FSA decisions has been limited. The effects of these tighter restrictions will be to push marginal operations out of farming altogether and to speed up the process through which their former lands are redistributed to larger, more successful farm businesses.

Thus, due to the pattern of contract payments distribution, the restructuring of credit relations, and the implementation of bare market competition, the FAIR Act will make it more and more difficult for family farms to compete and in this way will accelerate the historic trend toward farm consolidation.

As the framers of the Freedom to Farm concept hasten to point out, this represents a basic shift in the intent of farm legislation as updated over the past 60 years. True enough. Historically, U.S. farm policy has had a dual character. A key motive behind government intervention in agriculture was to stabilize rural society—to cushion farmers and therefore rural communities from the harsh production and price cycles of the market by providing a floor through which farmers could not fall. But there was a broader economic-development motive as well. By smoothing the cyclical nature of agriculture, price and income supports allowed farmers to plan and, more importantly, to expand production through investment in land, buildings, machinery, and a host of other manufactured inputs. Federal involvement in agriculture (through commodity programs and the system of farm research and information extension) thus acted to deepen the link between farming and agro-industries of all kinds, extending a process of agro-industrialization based on the expansion of the division of labor surrounding farm production (Goodman, Sorj, & Wilkinson, 1987; Page & Walker, 1991). So too, this focus on production and surplus pushed farmers into a model of technological competition. In order to take advantage of farm programs and market opportunities, farmers eagerly adopted a host of new farm techniques that boosted productivity. Yet as new techniques were generalized, production increased, prices fell, and many farmers were weeded out (Cochrane, 1979; Strange, 1986).

State management of agricultural prices was based on the generation, storage, and disposal of surplus commodities, which required, in turn, the subsidization of export subsidies and the active creation of foreign market (or giveaway) outlets (Friedmann, 1993). But since the 1970s, this system of surplus disposal has become more and more difficult to manage. Ironically, the chief problem for U.S. policy makers has been the replication of surplus in other parts of the world due to the spread of the American model of state-led agricultural development (a model that was actively promoted by the U.S. government in the postwar era). Global overproduction, particularly in grain crops, has led to declines in market prices and has thus increased the cost of maintaining prices for U.S. farmers, leading to a crisis in agricultural policy. In addition, competition between the agricultural sectors of various nations has led to heightened trade tensions and strained diplomatic relations.

The FAIR Act is a response to this predicament. It cleaves the traditional dualism of U.S. farm policy. In essence, the federal government has been forced to abandon the goal of rural social stability due to the very success of its productivist policies worldwide. The United States can no longer afford the high (if variable) cost of traditional farm programs or justify these programs politically at home or abroad. Many have argued that this course of action is warranted because the household income of commercial farmers now exceeds that of nonfarmers; in other words, the basic need to stabilize and improve farm income no longer exists. The problem with this assessment is that it is based on aggregate farm income figures that mask the differential fortunes of various types and sizes of farm operations. The removal of stabilizing farm programs will intensify swings in market prices and accelerate the workings of the technological treadmill. The only possible result is further farm consolidation and industrialization and the loss of more family farms.

■ Conclusion

If family farms are the losers in the restructuring of U.S. farm policy, then the question of the winners remains. Who really benefits from this historic shift in agricultural policy? Once the state withdraws, who will regulate agriculture? That is, who will organize stable relations of production and consumption? Who will dominate and direct the market? The answer to these questions is agro-industry, the most powerful set of actors in the farm economy.

Indeed, the agribusiness community was a chief sponsor and supporter of the FAIR Act. This position, however, represented a departure from the political alliances that had supported farm policy for 60 years.

In the past, both the farm community and agro-industries benefited from the federal management of agriculture, and together they comprised a powerful political force in Washington. But the interests of these two broad constituencies no longer coincide. The agro-industrial firms that were nurtured by productivist farm policy (e.g., seed producers, chemical input manufacturers, grain traders, food processors) now have a very different agenda, born of an expanded geographical scope. They no longer need the farm support component to ensure either an adequate supply of farm commodities or an adequate market for their manufactured farm inputs. Many (though certainly not all) agro-industrial firms can now source and sell globally, and in this way they have an ability to breech the confines of national policy measures (Friedmann & McMichael, 1989; Heffernan & Constance, 1994; McMichael & Myhre, 1991).[7]

The FAIR Act aids agro-industry in a variety of ways. Market transition will encourage an expansion of crop production (at least in the short term), and many firms will thrive on fencerow to fencerow planting—more acreage means a larger market for seeds, fertilizer, pesticides, herbicide, and so on. Increased production combined with farm severance payments will likely spark a rash of sales of new farm machinery. More crops also translates into higher volumes of commodities moving to market and thus increased earnings for commodity brokers, storage facilities, and shippers. Agro-industries will also benefit from the overproduction that soon will follow. Surplus production and subsequent price declines will lower the cost of farm commodities to those industries that use them as raw material—for instance livestock feeders, oil pressers, and food manufacturers of various types. While it sheds one half of traditional farm policy, the FAIR Act maintains the other through the promotion of world trade and exports, particularly value-added (manufactured) exports—programs designed to aid agro-industrial firms in their search for new markets. Finally, the global liberalization of trade in agricultural commodities and food products works to the advantage of several key agro-industrial firms by allowing them greater leeway in organizing the global integration of agro-food systems.

It is perhaps too soon to cast these changes as the triumph of international capital over the nation-state (e.g., McMichael & Myhre, 1991). We are not necessarily entering an era of private corporate regulation of agriculture in the context of open world trade. Such an interpretation underestimates the lasting importance of the nation-state by overestimating the stability of the current domestic political alignments and resultant agricultural policy. It also underplays the continuing ability of nation-states to circumvent trade rules, overlooks the importance of regional trading blocs, and ignores the subnational regulation of agriculture (cf. Bonanno, Constance, & Hendrickson, 1995; Goodman & Watts, 1994). Indeed, the restructuring of agricultural production in the United States involves struggles over governance occurring at a variety of geographical scales. In hog production, for example, the key issues are being decided in statehouses and in the meeting rooms of county supervisors (Page, 1996). Nor should we assume that all sectors of agro-industry have the same set of interests. As Goodman et al. (1987) point out, conflict among firms and industries with competing market strategies and different regulatory agendas can be fierce.

Nonetheless, we are witnessing an historic shift in farm policy, one that favors agro-industries overall. Despite this, farmers supported the FAIR Act overwhelmingly. Why? First, high severance payments were something they simply could not refuse. Second, farmers believed the prevailing wisdom that market transition will ultimately work to their advantage by granting increased autonomy in planting decisions, and that the continuing subsidization of exports will work to offset severe fluctuations in farm price levels. The problem is that price declines (which will occur at some point) never benefit farmers but always benefit some sectors of agro-industry. Moreover, even when times are good, farmers do not share equally in the expansion of agricultural production and markets. The gap between farm and retail prices has widened continually in the postwar era; overall, the percentage of the final value of food and fiber commodities received by farmers dropped by 11% between 1984 and 1994 and now stands at just 24%. The remainder goes to processors, shippers, wholesalers, and retailers (USDA, 1996d). Meanwhile, the cost of necessary farm inputs has risen steadily—placing farmers in the vise of a relentless price-cost squeeze (Commins, 1990).

A third reason that farmers supported the FAIR Act is that the issue of rural social stability never really entered the debate. Compared to the extensive discussion of the market transition program, new farm credit policies that make it more difficult for family farms to survive or to get back on their feet received very little attention. In fact, the entire question of a social safety net for farmers—in the form of credit programs, commodity programs, and so forth—was sidestepped. The manner in which this occurred may have important implications for future legislation. Key Democratic members of the House and Senate who in the past might have battled over this issue took a different approach. Realizing the political force behind the elimination of farm supports, they drew the line at the environmental provisions of the Farm Bill. This effort successfully preserved most of the environmental protections of previous farm legislation and even extended them a bit (Cushman, 1996). What this amounted to was a strategic trade-off between the issues of rural social sustainability and environmental sustainability.

The Clinton administration followed a similar path. Realizing that a strong stance on the environment was good for public relations during the budget battles of the fall of 1995, the administration pushed for the maintenance of conservation programs while demanding more funds for rural development. In addition, the retention of the 1949 law as permanent legislation also represented something of a victory for the administration (Jones, 1996a). But if the price for White House acceptance of the market transition program was the Fund for Rural America, this represents a curious disconnection between agriculture and rural development—an approach to rural society that weakens, rather than strengthens, the position of small and mid-sized family farms. In much of rural America, these farm families have been the backbone of rural economies; they purchase farm input and household goods locally and are active participants in community life. The Fund for Rural America will have to fight upstream to establish economic opportunity in rural places that have experienced population loss on the basis of farm consolidation and are witnessing the expansion of corporate, often absentee-owned farm operations that are dependent on low wage, nonunionized rural laborers. The long-term goal of environmental sustainability may be all the more difficult to achieve in this social and economic context. Here we encounter a final irony. For all the talk about the need for state devolution and greater levels of local participation in governance, the FAIR Act is a model of legislation that bears the unmistakable imprint of globally oriented business, is implemented at the federal scale, and excludes the voices of the rural people who are most affected by its policies.

As Congress sets out to remake much of the landscape of U.S. social and economic policy, one crucial question that emerges is whether the trade-offs apparent in the remaking of farm policy reappear in other policy arenas, or if the goals of social equity and social justice can be recombined with a strong commitment to the environment. This may

FAIR or Foul?

yet occur in agriculture. Overlooked this time around, the enhancement of rural social stability through the refinement of farm programs is an issue that will most likely reappear on the political agenda once the coming rural crisis has unfolded.

NOTES

1. This analysis of new farm policy is derived from a reading of the legislation as well as several USDA summaries (USDA, 1996b, 1996c).
2. Target prices go back to 1973, were originally tied to production prices, and were available only to those that complied with ARP requirements. Since 1981, target prices have been set independently of production prices and have thus become a politically determined minimum income guarantee (Strange, 1986).
3. California is allowed a separate federal milk order. In addition, New England may enter into a Dairy Compact if the Secretary of Agriculture identifies a compelling public interest to do so.
4. Under a TRQ system, a certain amount of imports—the quota amount—receives a low tariff, while imported amounts above the quota level are assessed a higher tariff.
5. It must be pointed out, however, that within production contract systems, contract growers are usually small-scale and ostensibly independent. This sort of operation would qualify for EQIP benefits, but assistance of this kind would also benefit the larger firms that are tied to growers within the production system.
6. However, rice appears to be the one component of Southern farming that stands apart from the trend. Southern senators successfully gained a payment "sweetener" for rice growers, but the elimination of price and income supports in rice production is projected to cause adjustments so severe that many farmers are expected to switch crops entirely in order to stay in farming (Smith et al., 1996).
7. This disjuncture between national interests and corporate interests was evident during the Carter administration's grain embargo of the Soviet Union. In that case, American grain-trading firms pursued their own business interests by running grain into the Soviet Union, circumventing the embargo and weakening its overall effectiveness (Friedmann, 1993).

REFERENCES

Bonanno, A., Constance, D., & Hendrickson, M. (1995). Global agro-food corporations and the state. *Rural Sociology, 60,* 274-296.

Cochrane, W. W. (1979). *The development of American agriculture: A historical analysis.* Minneapolis: University of Minnesota Press.

Commins, P. (1990). Restructuring agriculture in advanced societies: Transformation, crisis and response. In T. Marsden, P. Lowe, & S. Whatmore (Eds.), *Rural restructuring: Global processes and their responses* (pp. 45-76). London: David Fulton.

Cushman, J. H. (1996, March 19). Battle lines are being set up on sweeping environmental provisions in farm measure. *New York Times,* A14.
Environmental Working Group. (1996). *Freedom to farm.* Washington, DC: Author.
Food and Agriculture Policy Research Institute. (1996). Net farm income called optimistic by new FAPRI 10-year projection. Press release, FAPRI, University of Missouri, Columbia.
Friedmann, H. (1993). The political economy of food: A global crisis. *New Left Review, 197,* 29-57.
Friedmann, H., & McMichael, P. (1989). Agriculture and the state system. *Sociologia Ruralis, 29,* 93-117.
Goodman, D., Sorj, B., & Wilkinson, J. (1987). *From farming to biotechnology: A theory of agro-industrial development.* Oxford, UK: Basil Blackwell.
Goodman, D., & Watts, M. (1994). Reconfiguring the rural or fording the divide: Capitalist restructuring and the global agro-food system. *Journal of Peasant Studies, 22,* 1-49.
Heffernan, W. D., & Constance, D. (1994). Transnational corporations and the globalization of the food system. In A. Bonanno, L. Busch, W. Friedland, L. Gouveia, & E. Mingione (Eds.), *From Columbus to ConAgra: The globalization of agriculture and food.* Lawrence: University Press of Kansas.
Ingersoll, B. (1996, April 1). Congress passes new farm bill that dismantles subsidy programs after much vote trading. *Wall Street Journal,* A16.
Jones, H. (1996a, April 1). House receded to much of Senate farm bill during conference committee. *Feedstuffs,* 2.
Jones, H. (1996b, June 10). Ag spending challenges house committee. *Feedstuffs,* 1, 8.
McMichael, P., & Myhre, D. (1991). Global regulation vs. the nation-state: Agro-food systems and the new politics of capital. *Capital and Class, 43,* 83-104.
Page, B. (1996). Across the great divide: Agriculture and industrial geography. *Economic Geography, 72*(4), 376-397.
Page, B., & Walker, R. (1991). From settlement to fordism: The agro-industrial revolution in the American Midwest. *Economic Geography, 67,* 281-315.
Smith, E. G., Richardson, J. W., Gray, A. W., Klose, S. L., Outlaw, J. L., Miller, D. W., Knutson, R. D., & Schwart, R. B. (1996). *Representative farms economic outlook: FAPRI/AFPC April 1996 baseline.* (AFPC Working Paper 96-1). College Station: Texas A&M University, Department of Agricultural Economics, Agriculture and Food Policy Center.
Strange, M. (1986). *Family farming: A new economic vision.* Lincoln: University of Nebraska Press.
USDA. (1996a). *Proceedings, agricultural outlook forum '96.* February 21 & 22, 1996. Washington, DC: USDA.
USDA. (1996b). *The Federal Agriculture Improvement and Reform Act of 1996: Title-by-title summary of major provisions of the bill.* Washington, DC: Office of Communications, United States Department of Agriculture.
USDA. (1996c). *1996 FAIR Act frames farm policy for 7 years.* Washington, DC: Economic Research Service, United States Department of Agriculture.
USDA. (1996d). Agricultural fact book, 1996. Washington, DC: Office of Communications.

9

Back to the Future in Labor Relations: From the New Deal to Newt's Deal[1]

ANDREW HEROD

In this chapter, I examine a number of current issues concerning neoliberal proposals to change the manner of state regulation of the paid workplace. In the United States, two bodies of law relate directly to the workplace—*labor law*, which lays out rules for collective bargaining to be conducted *privately* between management and labor, and *employment law*, which ensures certain *public* rights for workers regardless of whether they are covered by a collective bargaining agreement. Across the political spectrum in recent years, questions have arisen as to whether the current system of labor and employment law is sufficiently suited to maintaining U.S. competitiveness and to protecting workers' rights in a rapidly globalizing economy, and whether the new realities of the U.S. labor market and workforce (such as the tremendous growth of "flexible" production techniques) require restructuring the current system of labor relations' regulation. I examine this debate as it has played out during the 1990s by analyzing recommendations made for change by the Dunlop Commission (appointed by President Clinton to look into the state of worker-management relations) and the legislative agenda proposed by the Republican majority in the 104th Congress (1995-96).

Although there has been much disagreement, it is perhaps telling that there also seems to be a considerable degree of consensus between these two groups on a number of issues. In fact, the principle issue seems to be not so much the *direction* of change, but the *degree* of that change. Both the Clinton administration and the Republican majority have responded to growing global threats to U.S. competitiveness by adopting very similar neoliberal legislative agendas that advocate bringing about greater labor market flexibility by *privatizing* aspects of labor-management relations, *dismantling* certain regulations and regulatory bodies, and *devolving* to lower levels of government some of the responsibilities for workplace regulation.

This process is not unique to the United States. In his book *Work-Place: The Social Regulation of Labor Markets,* Jamie Peck (1996) argues that in many advanced capitalist countries global competitive pressures and the predominance of neoliberal discourse are encouraging the development of new regimes of labor control. In particular, while capital is continuing to *globalize,* there appears to be an increasing *localization* of much labor regulation. Although this is frequently portrayed as a way to make labor markets more flexible in an era of heightened global competition, the outcome of such localization, Peck suggests, is the emergence of a new regime of labor control in which "strategies to localize labor are being used to extend managerial dominance and state discipline" (p. 232).

In what follows, I first provide an outline of the system of labor relations that has developed during the past six decades. This provides a context for the second part of the chapter, which examines a number of recent proposals for change put forward by the Dunlop Commission and the Republican majority in the 104th Congress.

■ The Making of the Modern System of Labor Relations and Regulations

The beginning of the modern era of labor relations in the United States can be dated precisely. On March 23, 1932, Republican President Herbert Hoover signed into law the Norris-La Guardia Act (47 Stat.70), which marked a milestone in the development of the nation's system of labor relations.[2] Prior to the 1930s, the federal government rarely involved itself in the direct legislative regulation of the relationship between labor and capital. Passage of the Norris-La Guardia Act was historic because, for the first time, the federal government enacted legislation that would develop *at the national level* a new system of labor regulation designed to further the ability of workers to organize by outlawing "yellow-dog" contracts, by limiting the power of employers to use labor injunctions against workers and their unions, and by preventing the use of antitrust legislation to hinder union activities (Goldman, 1979). Following the election of Franklin Roosevelt, several additional pieces of federal labor law designed to allow workers to organize more easily were soon enacted (for an excellent survey of some of the principles underlying New Deal labor law, see van Wezel Stone, 1981). Of these, the most important was the 1935 National Labor Relations ("Wagner") Act (NLRA), which dramatically expanded the power and purview of the federal government in the arena of labor law. Specifically, the Wagner Act established the principle of "unfair labor practices" under which workers could challenge the illegal anti-union activities of their employers and set up the federal National Labor Relations Board (NLRB) to rule on the validity of such challenges. In addition, the Wagner Act gave workers the legal right to organize and bargain collectively through representatives of their own choosing and outlawed employer-dominated trade unions ("company unions"), a large number of which had been formed during the 1920s and early 1930s by employers intent on limiting genuine trade unionism.

The New Deal labor laws passed during the 1930s are important with regard to current debates about state restructuring for several reasons. First, this legislation was part of a larger Keynesian policy of demand management in which the state attempted to encourage mass consumption as a means to stimulate mass production and so ensure economic well-being. In the case of New Deal labor law, the goal of FDR's administration was to protect factory wages and make union formation easier as a way of increasing the amount of disposable income available to workers who would, so the theory went, increase their consumption of consumer durables and thus stimulate the economy. Second, the labor laws set in place during the New Deal were based on a notion that the government's role should be that of "impartial judge" between capital and labor, which were themselves seen to be in a conflictual relationship. Third, through bodies such as the NLRB, the government could bring greater public scrutiny to bear on what had traditionally been considered to be the private employer-worker relationship. Fourth, although the Wagner Act was *federal* legislation, the system of labor relations that it spawned was highly *decentralized* and *localized* in nature. The means whereby union "locals" could be formed were designed to pay greater attention to local than to national concerns in the belief (long held in the U.S. political psyche) that collective bargaining should be responsive to local conditions and needs of workers and employers, rather than to national or industry-wide conditions. Fifth, although New Dealers passed some federal legislation (such as the 1938 Fair Labor Standards Act) that mandated certain minimum wages and maximum hours, the model of liberal pluralism that guided their thinking placed greatest emphasis on a system of labor relations in which workers' rights would primarily be protected through collective bargaining.

The system of worker-management relations essentially functioned this way until the 1960s, with the federal government primarily seeking to regulate conditions in the workplace through *labor* law and the NLRB. However, in the wake of the Civil Rights movement, a dramatic change occurred as both federal and state governments increasingly enacted legislation related to *employment* law. Whereas New Deal legislation such as the NLRA had been an attempt to allow workers to be protected in the workplace through trade union activities and collective bargaining, employment law represented an attempt to ensure that workers' rights were protected instead through legislative fiat. Hence, the Equal Pay Act of 1963 limited gender-based differences in wages and benefits; the Civil Rights Act of 1964 limited discrimination in the areas of pay, hiring, firing, and promotion on the basis of race, sex, religion, or national origin; and the Age Discrimination in Employment Act of 1967 placed prohibitions on employment discrimination based on age. In the 1970s, this body of employment law was expanded by implementation of the Occupational Safety and Health Act of 1970, the Federal Mine Safety and Health Act of 1977, the Employee Retirement Income Security Act of 1974, and other pieces of legislation. In the 1980s, Congress passed several new employment laws regarding the hiring of illegal immigrants (the 1986 Immigration Reform and Control Act), notice of plant closures (the 1988 Worker Adjustment and Retraining Notification Act), workers with disabilities (the 1990 Americans with Disabilities Act), and family leave (the 1993 Family and Medical Leave Act). With such laws came several new regulatory agencies designed to enforce their provisions.

This change in the relationship between the state and the workplace is significant for five reasons. First, it greatly concentrated regulatory power in the hands of the federal and state governments. By so doing, it represented an explicit leveling of conditions at several geographical scales across the economic landscape, within both individual states and the nation as a whole. Although state laws could go further than federal minimum standards, in most cases they could not be allowed to go below them. This represented an implicit *nationalization* of standards affecting workplaces across the country, the imposition of a national bare minimum below which employers could not go unless given special exemption by federal legislative action. Certainly, the old Wagner Act may have been federal legislation, but it put in place a highly decentralized system of labor relations in which conditions could vary greatly across the country depending on the outcome of local collective bargaining. Federal employment law, on the other hand, greatly limited the options for local variability in workplace conditions by setting national standards. Second, under labor law it was the *conditions* for organizing and collective bargaining that were guaranteed at the federal level. Given the decentralized nature of the collective bargaining system, this ensured a wide variety of bargaining outcomes across the economic landscape as more powerful groups of workers were able to extract better conditions from their employers than could weaker groups. In contrast, the focus of employment law was on the *outcomes* of federal and state policy.

Third, the growth of employment law represented a further and much more widespread public intervention in the operation of the workplace than did the New Deal labor legislation. This is because employment law created *public* rights that could not be negotiated away privately through the collective bargaining process, and because the responsibility for interpretation and enforcement of such laws would now lie in the hands of various *public* officials such as government regulators and the courts, rather than the *private* representatives of management and labor. Fourth, as a consequence, employment law increasingly shifted much of the burden for enforcement of workplace rights out of the purview of the NLRB and into the courts as growing numbers of workers began to sue their employers for breaches of the law. Paradoxically, the protections afforded workers by their newfound *public* rights were increasingly to be guaranteed and enforced by litigation initiated and pursued *privately* by the individuals affected. Fifth, employment law covered a larger proportion of the workforce than did labor law, which covered only those workers (concentrated in the industrial heartlands of the Northeast and Midwest) who had unionized so that they might bargain collectively. This became increasingly significant as the proportion of the labor force that belonged to unions and was covered by collective bargaining agreements began to decline after the 1950s. In particular, the growth of employment law expanded geographically throughout the country many rights and protections that previously only unionized workers had enjoyed.

During the past three decades, then, a mode of government regulation in the area of worker-management relations has come to dominate that is different than that laid out under New Deal labor law. Whereas labor law essentially takes a hands-off approach to the outcome of collective bargaining, assuming that no laws have been broken, employment law imposes certain minimum standards and practices believed necessary by the government to ensure the health and welfare of workers. By so doing, federal and state governments have hoped to both prevent temptation on the part of managers and workers to give up such rights for the benefits that might accrue from noncompliance with these public standards and protect those employers who do follow such regulations from suffering the ill effects of unfair competition from those who might enjoy lower production costs through noncompliance.

The current economic and political situation in the United States is, in many ways, fundamentally different from that of the 1930s or even the 1960s. The shift from Keynesian demand-side economics toward the supply-side approach advocated by neoliberals; tremendous growth in the contingent (temporary and part-time) workforce; continued declines in union membership; greater global competition; the development of new flexible production technologies and systems, such as Computer Aided Design (CAD), Computer Aided Manufacture (CAM), and Just in Time (JIT) delivery; the tendency toward production at smaller plants; the growth of subcontracting; and the explosion of litigation under provisions of both employment and labor law have all stimulated an intense debate across the political spectrum about what should be the role of the federal and state governments with regard to regulating workplaces and labor markets. In this debate, both conservatives (as represented by the Republican agenda laid out in their capture of Congress in the 1994 elections) and liberals (as represented by President Clinton's appointment of a "Commission on the Future of Worker-Management Relations") have come in recent years to question whether the current role played by federal and state government is appropriate, given contemporary developments in the U.S. and world economy, and if not, what new role government should play, what institutions and regulations are necessary to enable it to play such a new role, and at what geographic level (national, state, or local community) such institutions and regulations should be constituted.

It is perhaps too early to make definitive statements with regard to how the state of labor relations is being restructured in light of such questions, given that much of the legislation proposed by the Republicans who went to Washington, D.C. in 1995 promising to overturn 60 years of "liberal welfarism" is still being developed, whereas many of the questions raised by the Dunlop Commission have yet to be acted upon. It is nevertheless significant that both groups have focused on some common issues, namely the "need" to reprivatize some of the responsibility for worker-management relations by limiting the scope of government intervention and the "need" to develop laws that are more sensitive to the variety of conditions found in workplaces across the country. The discourse on the subject from both groups is heavy with the language of "privatization," "deregulation," "labor market flexibility," "marketization," and the implementation of a "New Federalism."

■ Remaking the State and Labor Regulation in the 1990s

One of the first acts the Clinton administration took after coming to power in January 1993 was to initiate a wide-ranging investigation of the state of worker-management relations in the United States. Appointed by Secretary of Labor Robert Reich and Secretary of Commerce Ron Brown, and headed by John Dunlop (who had been Secretary of Labor under President Gerald Ford), the Commission on the Future of Worker-Management Relations (the "Dunlop Commission") held hearings in several cities across the country during 1993 and 1994. In particular, the Commission set itself three primary questions:

1. What (if any) new methods or institutions should be encouraged, or required, to enhance work-place productivity through labor-management cooperation and employee participation?
2. What (if any) changes should be made in the present legal framework and practices of collective bargaining to enhance cooperative behavior, improve productivity, and reduce conflict and delay?
3. What (if anything) should be done to increase the extent to which workplace problems are directly resolved by the parties themselves, rather than through recourse to state and federal courts and government regulatory bodies? (Dunlop Commission, 1994a, p. vii)

The assumption behind these questions was that because the U.S. economy has changed during the past six decades, the time has come to consider restructuring the mode of labor regulation to increase U.S. workers' competitiveness in the global marketplace.

Significantly, although the Republican majority in the 104th Congress may have differed from the administration in its political motives regarding labor-management relations and the role the federal government should play in these matters, much conservative rhetoric surrounding proposals to initiate a new model of workplace regulation parallels the administration's alarm calls about the growing competition U.S. workers and companies face from abroad. For example, the preamble to the "Teamwork for Employees and Managers Act of 1995" (HR 743, 104th Congress) introduced on January 30, 1995, suggested that "the escalating demands of *global* competition have compelled an increasing number of employers in the United States to make dramatic changes in workplace and employer-employee relationships" (emphasis added). Whatever the veracity of such claims, the articulation by neoliberals of a discourse of global threats to U.S. competitiveness is serving as a potent device to transform the nature of regulation to make labor markets more flexible.

The Dunlop Commission

In its study of the contemporary state of U.S. labor relations, the Dunlop Commission made several recommendations for revising regulations governing workers and their workplaces. One of the most significant of these is the encouragement of labor-management cooperation (for the sake of the "national interest"), which is to be achieved by facilitating the growth of programs allowing greater employee involvement in workplace decision making related to production, training, product quality control, safety, and health; worker representation on corporate boards of directors; and voluntary dispute resolution.[3] The model of employee involvement that many Commission members seemed to have in mind was that of the "works council," a body that, in many European countries, enables workers to participate in decision making at the individual plant level (see Thelen, 1991, for an excellent account of how works councils came about and operate in Germany). However, much of the activity conducted by such works councils would be deemed illegal under Section 8(a)(2) of the current NLRA, which makes it an unfair labor practice for an employer to "dominate or interfere with the formation or administration of any labor organization or to contribute financial or other support to it," with *labor organization* being defined very broadly under the Act as "any organization of any kind, or any agency or employee representation committee or plan, in which employees participate and which exists for the purpose, in whole or in part, of dealing with employers concerning grievances, labor disputes, wages, rates of pay, hours of employment, or conditions of work." Consequently, although the Commission was keen to "reaffir[m] the basic principle that employer-sponsored programs should not substitute for independent unions . . . [or] be legally permitted to deal with the full scope of issues normally covered by collective bargaining" (Dunlop Commission, 1994b, p. 8), it did advocate amending the NLRA to clarify the legal status of employee involvement in workplace decision making, which has, in any case, grown in recent years through the spread of activities such as self-managed work teams, quality circles, and joint labor-management committees, but which has been challenged in the courts under Section 8(a)(2).

Following quickly on the Commission's heels, in January 1995, the Republican-controlled Congress introduced bills in the House (HR 743) and Senate (S 295) proposing the "Teamwork for Employees and Managers Act" (commonly referred to as the TEAM Act) that would amend the NLRA precisely to allow such workplace "employee involvement programs." Although the TEAM Act bills were ultimately vetoed on July 30, 1996, by President Clinton, they are nevertheless indicative of the way in which conservatives have attempted to remake U.S. labor relations. (Significantly, in his veto message Clinton did not oppose the notion of labor-management cooperation and teamwork per se, but argued instead that this particular bill would not promote what he called "*genuine* teamwork"). In particular, the bill would have amended the NLRA to allow employers to "establish, assist, maintain, or participate in any organization . . . in which employees . . . participate to at least the same extent practicable as representatives of management . . . [so that labor and management might] address matters of mutual interest," such as issues of product quality control, productivity, efficiency, and health and safety. The bill's proponents argued it would allow "legitimate employee involvement programs, in which workers may discuss issues involving terms and conditions of employment, to continue to evolve and proliferate" (HR 743), thereby encouraging labor-management cooperation and so reducing strikes, increasing U.S. competitiveness, and improving workers' work experiences. They also maintained that such legislation would not impinge on unions' abilities to represent workers and to bargain collectively if workers so wished, because employee involvement programs would not deal with issues such as wages, nor would the Act allow company unions to be set up. Opponents, however, maintained that such legislation would make it increasingly difficult for many unions to represent workers, because such programs could perhaps be instituted by employers and compliant workers to stave off the threat of a union-organizing drive. Furthermore, they argued, joint labor-management committees not linked to a trade union also may increasingly come to rival unions as representatives of workers' interests, particularly given the low levels of unionization in the United States relative to countries such as Germany, where worker representatives on the works council are more often than not union members.[4]

The recommendations of the Dunlop Commission and the provisions of the TEAM Act are significant for a number of reasons. First, both represent a dramatic shift away from the model of labor relations on which labor law has been based since the 1930s in which capital and labor were assumed to have an adversarial relationship and in which the role of government (in the form of the NLRB) was to act as referee between the two. The new model would be one that, in the words of the Committee on Economic and Educational Opportunities (House Report 104-248), "fosters cooperation, not confrontation" between management and labor, and one in which the distinction between worker and supervisor is reduced as workers take on more responsibilities typically thought of as managerial.[5] Such a model is consistent with "flexible"/"Japanese" forms of production organization and workplace relations that have few skill classifications and in which workers may rotate through several types of jobs as members of "teams," a model that is being adopted by many firms as they seek to emulate companies such as Toyota and Honda. Second, such recommendations represent an effort to *decentralize* labor-management relations by increasing the significance of employee participation and involvement at the level of individual plants.[6] Some observers argue that this may dramatically increase "plant chauvinism" as workers show greater allegiance to their own particular plants than to workers in other plants, thereby reducing the potential for solidarity across the economic landscape during disputes or collective bargaining—a prospect no doubt welcomed by neoliberal advocates of labor market "flexibility."

Third, while employee involvement plans may on the one hand allow workers and employers to negotiate arrangements that give them greater flexibility to respond to local conditions and concerns, they also significantly increase the potential for greatly different types of participation programs to be developed in workplaces across the country. How workers and employers choose to deal with particular issues in their own plants may vary considerably. In turn, this provides greater opportunity for mobile capital to play different plants off against each other on the basis of the minutest differences in their in-house programs and local regimes of labor relations. The economic landscape, in other words, would be opened up to the forces of competition to a much greater degree than before. Fourth, the recommendations also represent an implicit *privatization* of the labor-management relationship, because they limit the power of governmental agencies, such as the NLRB, to regulate the kinds of arrangements workers and managers may develop voluntarily. Indeed, such was explicitly stated in the preamble to HR 743, which argued the Act would "*protect* legitimate employee involvement programs *against governmental interference*" (emphasis added). By reducing what is illegal, the Act implicitly gives workers and management a greater variety of potential arrangements from which to choose.

With regard to employment law, both the Commission and the Republican majority in Congress have suggested that the myriad federal regulations enforced by many different agencies make it difficult for both workers and employers to know their legal rights. These regulations, they claim, also limit flexibility because they take nationally developed standards and apply them locally. Consequently, both Congress and the Commission have recommended a move toward *devolving* some responsibility for regulation and *privatizing* much of the activity relating to workplace dispute resolution currently conducted by governmental agencies and bodies. Noting that it "is increasingly difficult to write and enforce standard regulations that fit well with the diverse employment settings and workforce and the changing workplace practices found in the contemporary economy" (Dunlop Commission, 1994a, p. 108), the Commission strongly "support[ed] the expansion and development of alternative workplace dispute resolution mechanisms, including both in-house settlement procedures and voluntary arbitration systems that meet specified standards of fairness" (Dunlop Commission, 1994b, p. 27).

Three types of private alternative dispute resolution (ADR) are typically available. These are (a) a system in which the parties may attempt to resolve their dispute voluntarily through direct, in-house negotiations; (b) a system in which the parties turn to a mediator sponsored by the courts or government, whose decision both accept voluntarily; and (c) a system in which a legally binding decision is made through use of an arbiter (Dunlop Commission, 1994b, p. 26). Both the Commission and the U.S. Supreme Court have shown themselves receptive to the use of ADR as a way to cut down on the litigation that has arisen in the wake of employment law. Although in some ways ADR could be seen merely to represent a shift from using private litigation in employment law dispute resolution to using private mediation/arbitration, in other ways it can be seen actually to represent a qualitatively *greater* privatization of workplace dispute resolution because it replaces government regulators with private/nongovernment agents. Furthermore, the prospect of going to arbitration may, in fact, encourage recalcitrant parties to reach mutually acceptable compromises privately before the NLRB or courts step in, thereby "[m]aximizing the number of such voluntary agreements [which] is the goal of any dispute resolution system" (Dunlop Commission, 1994b, p. 22). The Commission's approach, it seems, is one in which within the framework of *national* legislation, *private* arbitration that is sensitive to the needs of *local* collective bargaining is to be encouraged as a way of solving conflicts *voluntarily* in the private arena. Only as a last resort would federal agencies be called on to make determinations concerning workplace disputes. Although to date only the Equal Employment Opportunities Commission and the Department of Labor have experimented much with ADR, the growing cost and quantity of litigation mean that it will probably expand in the future, particularly given that the alternative is to increase the budgets of various agencies so that they may handle the increased caseloads—something which, in an era of "government downsizing," lacks sufficient political support.

Related to the issue of developing workplace regulations to reduce disputes voluntarily and without recourse to government agencies, the Commission has advocated expanding the practice of "negotiated rulemaking" ("reg-neg") in which the various parties who might be affected by a potential government regulation (including the government agency itself) are encouraged to reach some consensus before the rule is published for public comment. Reg-neg is a voluntary process in which the participants establish their own procedural rules and in which an impartial mediator seeks to facilitate discussion. It is designed to produce regulations that are less likely to be challenged in litigation. Congress endorsed the notion of reg-neg in its 1990 Negotiated Rulemaking Act (Public Law 101-648), and in October 1994, Secretary of Labor Robert Reich announced that all of his department's agencies would consider using reg-neg in future regulations' designs. The Commission's endorsement of reg-neg as a means to reduce litigation under employment law is significant for at least two reasons. First, by allowing the affected parties to negotiate their own procedural rules and to participate in the formulation of regulations, reg-neg represents an explicit *devolution* of responsibility for regulating U.S. workplaces. No longer would rules simply be established by federal government agencies and imposed unilaterally from Washington, DC. Instead, individual regulatory agencies would be encouraged to "develop guidelines for internal responsibility systems in which parties at the workplace are allowed to apply regulations to their circumstances" (Dunlop Commission, 1994b, p. xix). Hence, affected parties from different industries and different parts of the country would now be directly involved in regulation formulation, a practice that could result in standards being developed that more accurately reflect local or regional conditions and needs than national ones. Equally, depending on how the process of reg-neg might operate, allowing workers and management in particular workplaces to develop individualized regulations could lead to a more variegated regulatory landscape.

Second, by shifting responsibility for the development and enforcement of regulations affecting workplaces to workers and management, the Dunlop Commission (1994b) hoped that such a system would "resolve more regulatory disputes in the workplace itself, while still providing the employee with recourse to a neutral agency where satisfactory resolution of the problem on the job is not forthcoming" (p. 50). In other words, the system of reg-neg would represent a greater *privatization* of labor relations' regulation by encouraging labor and management themselves to resolve regulatory issues directly and to use the particular agency under which the regulation falls as a neutral arbiter/enforcer of last resort. Firms would respect such rules, the Commission reasoned, because failure to do so would leave them open to (*privately*) initiated) litigation.

The Republican Agenda

Efforts by the Republican majority in Congress to restructure regulations concerning labor relations have likewise focused on issues of *privatization,* the *dismantling* of government regulatory agencies, and the *devolution* of responsibility for regulation from the federal government. As with the recommendations of the Dunlop Commission, many of these proposals have yet to be fully enacted. Nevertheless, they do give a sense of the recent trend in government and conservative public policy circles and how this is likely to play out in the future.

Perhaps the most frequently articulated goal of the Republican majority is that of reducing government regulation of workplaces so that market forces of supply and demand can play a greater role in determining the outcomes of decisions made in the workplace privately by what are, conservative economic theory holds, millions of atomistic individuals. This is to be achieved by dramatically scaling back the regulatory power of the federal government in a process that would reduce *public* oversight of many workplaces, *devolve* to state and local agencies the responsibility for the enforcement of rules, and *dismantle* other agencies and employment laws. One of the most far-reaching bills intended to facilitate this was one introduced in June 1995 by Representative Cass Ballenger (R., NC). Under his "Safety and Health Improvement and Regulatory Reform Act" (HR 1834), the 1970 Occupational Safety and Health Act would be amended to require the setting and revision of OSHA standards to be based on certain analyses and criteria, including a specified type of regulatory impact and cost-benefit analysis to ensure that any standard is "economically feasible."[7] The bill emphasized warning, rather than fining, employers not in compliance with the regulations. It also favored changing reporting procedures so that an employee would first be required to report a hazard to the employer (raising the issue that such employees may open themselves to being fired as "troublemakers") and then to prove that no improvement in the situation had been made before filing a complaint with OSHA. By advocating voluntary solutions between workers and employers, the bill would greatly *privatize* the worker-employer relationship through delaying recourse to OSHA and by shifting much of the burden for regulatory enforcement from federal government inspectors to individual employees in particular establishments.

In addition to these provisions, the bill proposed eliminating requests for inspections by employee representatives (e.g., unions) and making inspections discretionary rather than mandatory. Furthermore, it would exempt from OSHA inspections workplaces employing fewer than 50 workers unless they have more injuries than the national average for their industry, while also excluding from regulation any firm/person engaged in a farming operation with fewer than 10 employees and that does not have a temporary labor camp.[8] Such provisions would effectively exempt about 75% of all workplaces and would therefore represent a massive *privatization* of workplaces in which millions of workers would, to all intents, be denied the protections of *public* federal employment law. The bill also proposed to *dismantle* several regulations concerning air quality, chemical inventory reporting, and ergonomic standards. As part of the effort to reduce the federal government's influence in the workplace, the bill also provided for prosecution under state and local criminal laws, a *devolution* of regulatory responsibility. Finally, the bill proposed repeal of the Mine Safety and Health Act and the *dismantling* of the Mine Safety and Health Administration, the reduction of mine inspections from four times to once per year (a further *privatization* of the workplace), elimination of surprise inspections, and limiting the right of miners to take cases to court (a *dismantling* of their *public* rights previously guaranteed under employment law).

The Senate version (S 1423) of the bill, though generally seen as less drastic, contained many similar provisions (e.g., repealing the requirement that OSHA conduct an inspection on receipt of written complaint) and proposed exempting firms from OSHA health and safety inspections based on reviews by state consultation programs, insurance carriers, or other third parties that certified that the firm could correct the problem (again, a *devolution* and *privatization* of responsibilities previously carried out by the federal government).[9]

Significantly, while the Republican majority's rhetoric maintains that its goal is to bring about a greater play of market forces in the economy and return power to states and local communities, this goal is not, in fact, to be achieved solely through *privatizing* the workplace, *devolving* responsibilities, and *dismantling* agencies. Indeed, for all the talk of a Reagan-inspired New Federalism and ending "big government," in some cases Republican strategy has actually involved expanding the power of the federal government to so burden the regulatory system with government oversight as to make regulation practically impossible. In arguably the ultimate example of political chicanery, *deregulation* and *privatization* are actually to be achieved through *overregulation* and increased *public* intervention by the federal government. Hence, Ballenger's bill (HR 1834) allows any person affected by a standard to petition the Secretary of Health and Human Services to conduct a cost-benefit analysis of any standard, a proposal guaranteed to delay implementation of regulations.

Likewise, in the Contract With America Advancement Act (PL 104-121) passed in March 1996, Title II of the Act (known as the "Small Business Regulatory Enforcement Fairness Act") amends the Equal Access to Justice Act to require OSHA to pay the legal fees of firms or individuals who challenge the agency's "excessive" enforcement activities, while it amends the Regulatory Flexibility Act to allow greater challenge of OSHA standards by small businesses. In an attempt to undermine OSHA's regulatory ability, the Act *devolves* to a number of "Small Business Regulatory Fairness Boards" (to be established in each of the Small Business Administration's [SBA's] regions and made up solely of representatives of small businesses) the power to complain to a "Small Business and Agriculture Regulatory Enforcement Ombudsman" while shifting *horizontally* within the federal government regulatory responsibility by giving the SBA substantial oversight over OSHA inspectors and activities. The Act also sets up a process by which Congress can review and veto all OSHA regulations, thereby shifting power from the Executive to the Legislative branch of government and opening the way for special interest lobbyists to have much greater influence on regulatory enforcement activities. Paradoxically, the greater *privatization* and *devolution* of regulatory responsibility so valued by conservatives in their vilification of big government is actually to be achieved through greater *public* intervention on the part of Congress. Such examples remind us that the much-talked-about "downsizing" of government is, in fact, partial; for in many cases, big government is precisely what the neoliberal Right desires to achieve its aims.

■ Conclusion

Clearly, it is not possible to examine every recommendation or piece of legislation proposed during the past few years that has aimed to remake the system of labor relations in the United States. Instead, I have selected a few significant examples to show how these would fundamentally transform the nature of government and regulation. Although not all have been fully implemented, they are, I believe, indications of what the future is likely to bring. This is particularly so, given the current political climate and the fact that many liberals and conservatives in government seem to be in general agreement about the direction that the system of worker-management relations should take in response to global competition and the shift to more flexible modes of work (Quality of Worklife Circles, teamwork, the growth of flextime, and contingent workers), which some have heralded as characteristic of a "postfordist" economy. Hence, whereas for conservatives, state restructuring and the localization of labor market and workplace regulation is seen as a way to stimulate competition and protect free market forces from distortion by government, for many liberals and radicals it is viewed as a way of allowing greater worker and employer flexibility to respond to local conditions and needs.

As we have seen, the primary purpose of much of the current effort to restructure worker-management relations and the laws governing them has been to replace public rights guaranteed centrally through federal law with private rights established in the marketplace through "voluntary" negotiations between workers and management and with more geographically circumscribed public rights guaranteed by state or local law, rather than by federal law. Whereas workplace regulations developed under fordism sought to establish nationally a regulatory *floor* below which minimum standards could not fall, under a post-fordist and/or neoliberal regime of labor market and workplace regulation, it is increasingly a *ceiling* that is being imposed above which individual states and communities will not de facto be able to go for fear of losing capital investment to other, less-regulated states and communities (cf. Peck, 1996).

Such a transformation in the regulatory regime has a number of implications. First, it will dramatically reduce the protections that many workers enjoy, because it will cover only those who are sufficiently powerful to negotiate voluntarily agreements containing provisions previously guaranteed by law. Second, it will result in a much more variegated economic, social, and political landscape as regulations are developed and implemented locally rather than nationally. Although this may allow more powerful groups of workers to negotiate conditions above the national average, for others it will undoubtedly lead to even greater pressure to try to undercut competitors in a "race to the bottom" as states and workers make concessions on issues related to, say, health and safety or environmental pollution in an attempt to attract more mobile investment. Implicitly, this will, in fact, also undermine the ability even of well-organized workers to negotiate strong protections for fear that mobile capital may leave their communities for less-regulated pastures. In sum, it will result in a landscape that increasingly reflects the power of the market to shape economic, political, and social geographies while diminishing the power of government to do so. Given the nature of how markets operate under capitalism, this will be a landscape of increasing inequality, in which the outcomes of competition are writ large, with disparities of wealth, opportunity, health, and safety becoming greater as the powerful are able to shape conditions to their own benefit and the weak are left behind.

The localization of labor market and workplace regulation, then, raises important political questions for progressives, particularly given that it has become fashionable to celebrate the local as being somehow inherently "democratic" because it allows sensitivity to difference and local expression, while decrying the nonlocal as "totalitarian" and "oppressive" because it supposedly subsumes difference and imposes decisions made elsewhere in top-down fashion. However, localism is a double-edged sword that must be handled carefully if we are not to cut ourselves. Certainly, localism may provide the political space necessary for the celebration of difference and the development of a politics sensitive to this difference. But it can just as easily open the way to much greater competition between places and allow to flourish exclusionary localist ideologies that are manifested in ethnocentrism, anti-outsiderness, and local chauvinism.

Rather than privileging the local as more democratic simply because it is local, progressives must bear in mind that political struggle should always be multi-scale in nature. Local control may well serve in some instances as protection from repressive national regulations, yet it may also allow the undermining of policies that those of us on the left regard as important rights (such as nationally guaranteed abortion rights or civil rights). The issue is not whether certain scales of social regulation are inherently more or less liberatory or repressive, but of whose interests are served by regulation at particular scales. This is a historically and geographically specific question over which we must continually struggle.

NOTES

1. I would like to thank Jennifer Frum and Jan Kodras for comments on an earlier version of this paper.

2. Some might argue that the 1926 Railway Labor Act marks the beginning of the period of modern labor relations in the United States because it too was a piece of federal legislation designed to protect workers' rights to organize. However, this Act was much more narrowly circumscribed in extent in that it was applicable only to railway workers (although in 1936 it was amended to expand coverage to the airline industry as well), rather than to the broader mass of workers affected by the Norris-La Guardia Act. In

addition, it is quite ironic, given the vociferousness with which Newt Gingrich and other Republicans have railed against six decades of "big government" and "federal interference," that the first forays into establishing a system of federal regulation of labor relations occurred under the watch of Republican Presidents Coolidge and Hoover.

3. As the AFL-CIO (1994) noted in its Executive Committee Report, however, "employee participation" means different things to different people. In particular, the Report draws a distinction between "worker empowerment," which is task driven and in which employers are willing to give workers more control over the work process because they realize it is a better way to get the work done, and "workplace partnership schemes," in which workers have some power to determine the terms and conditions of employment under which they labor.

4. Interestingly, this is also a concern of some Eastern European trade unionists in those countries where there has been talk of introducing German-style works councils. Jan Uhlíř, the national president of the Czech Metalworkers' Federation (Odborový Svaz Kovo), has stated that his union fears that some employers may seek to establish works councils as a way of forestalling dealing with the union (personal interview, August 24, 1994).

5. This does not mean to say that the Commission has sought to blur the line completely between workers and managers. Indeed, the Commission was quite clear to state that true managers should continue to be excluded from the provisions of the NLRA. However, it recommended attempting to clarify the law so as to allow those workers who participate in some kinds of decision making to bargain collectively if they so desire. In this sense, the Commission (1994b, pp. 9-10) attempted to draw a distinction between "managerial employees" (statutory supervisors and managers) and members of workteams or professionals to whom authority is sometimes delegated so that they may direct their less skilled co-workers.

6. The House Committee on Economic and Educational Opportunities explicitly recognized this goal, stating "The Committee has focused several of its legislative efforts on *decentralizing* decisionmaking in a variety of areas within its jurisdiction. . . . [I]n the employment arena, employee involvement increases *local* decisionmaking by giving employees a voice in how their workplace is structured" (House Report 104-248, emphasis added).

7. In June 1995, the bill was referred to the House Committee on Economic and Educational Opportunities, and hearings were subsequently held by the Subcommittee on Workforce Protections. Under a follow-up bill (HR 3234), the Small Business OSHA Relief Act introduced by Ballenger in April 1996, any standard promulgated by OSHA would have to be "based upon an assessment of the costs and benefits of the standard and a determination that the benefits of the standard justify the costs imposed by the standard." HR 3234 was forwarded to the full committee on April 17, 1996, where it died.

8. Of course, with reduced regulation of safety codes, it would be expected that average injury rates would rise. It has been estimated, for instance, that between 1970 (the date of OSHA's creation) and 1995, workplace fatalities have been reduced by 55%, representing a saving of 140,000 lives.

9. The Senate version of the bill emerged from the Committee on Labor and Human Resources on June 28, 1996, and was then placed on the Senate legislative calendar. No further action was taken on the bill before the end of the Congress.

REFERENCES

AFL-CIO. (1994). *The new American workplace: A labor perspective.* Washington, DC: AFL-CIO Executive Council.

Dunlop Commission. (1994a). *Fact finding report: Commission on the future of worker-management relations.* Washington, DC: U.S. Department of Labor and U.S. Department of Commerce.

Dunlop Commission. (1994b). *Report and recommendations: Commission on the future of worker-management relations.* Washington, DC: U.S. Department of Labor and U.S. Department of Commerce.

Goldman, A. L. (1979). *Labor law and industrial relations in the United States of America.* Deventer, Netherlands: Kluwer.

Peck, J. (1996). *Work-place: The social regulation of labor markets.* New York: Guilford.

Thelen, K. A. (1991). *Union of parts: Labor politics in postwar Germany.* Ithaca, NY: Cornell University Press.

van Wezel Stone, K. (1981). The post-war paradigm in American labor law. *Yale Law Journal, 90*(7), 1509-1580.

10
Responsibility, Regulation, and Retrenchment: The End of Welfare?[1]

MEGHAN COPE

The most recent rounds of welfare "reform" enacted in the United States have again raised concerns at many levels, from local service providers to state and federal agencies, from welfare recipients to political elites. However, even a cursory look at "workfare" legislation, the devolution of social policy administration to states and localities, and the structural forces influencing the low-wage end of the labor market suggests strong links between the increasingly uncertain flows of international capital, the demise of fordism as the dominant production regime in the United States, and the rise of a public sentiment that sees federal social welfare programs as obsolete and undeserved.

In this chapter, I examine the ways that the project of state devolution has been expressed through welfare policy shifts. With the passage of the "Personal Responsibility and Work Opportunity Reconciliation Act" of 1996 (hereafter PRWOR), which ends the federal guarantee of support for poor families, shifts funding to state block grants, and strengthens time limits and work requirements, there is no question that devolution is having an impact on this area of social policy. But there are two important scales of analysis that need to be examined here, because the elimination of the federal guarantee will not merely result in 50 identical, state-level safety nets (Handler, 1995). Rather, both the broader structural developments that are accompanying the emergence of postfordist production at national and international levels and the dynamic regulation of labor markets within the United States come together to mandate a double-sided shift in, on one side, political rhetoric and public perceptions and, on the other, actual policy changes. Here we see that the concrete results of "competitive downgrading" (Peck, 1996, p.252) of states' and localities' social policies are creating a geographically uneven pattern of incentives, requirements, and sanctions that constitute a brutal new array of welfare policies. This fragmented system is indicative of what can be seen as the social policy corollary to the "flexible accumulation" of capital, that is, the flexible *dispersion* of labor.

■ Why Is Welfare a Target for Restructuring?

Despite its relatively tiny apportionment of the federal budget (approximately 1%), welfare—or the support of poor families with dependent children—has received disproportionate attention in recent years, both in terms of public sentiment and political rhetoric and in terms of actual policy changes. Although many social observers see this in terms of a continued "war against the poor" (Gans, 1995) or as the result of a misguided focus on the behavior of individuals (Handler, 1995), others suggest that the crises of *economic restructuring* are the fundamental points of departure for the renewed debate on social policy and the devolution project. In this framework, economic restructuring is seen as a transformation, occurring over the past 25 years, in the ways through which capital is accumulated, the speed at which it circulates the globe, and the degree to which it penetrates all regions of the world. Central to this transformation has been the globalization of manufacturing-based production and the consequent deindustrialization of the economy of the United States (and others). This, in turn, resulted in a shift away from the logic of the brief postwar heyday of fordism with its economic and social supports supplied by a Keynesian welfare state, toward something that has variously been called postfordism or postindustrialism. The question of what exactly this new condition or "regime" is—or if indeed we have reached something new yet—is still being worked out. What is clear is that there is a growing sense that in a country that is no longer primarily engaged in fordist production, there is no longer a need for the economic and social supports that made that regime possible, profitable, and palatable. Such is the way with welfare.

The welfare state was gradually constructed as a system of supports for those who—for one reason or another—were not expected, or were unable, to meet the demands of labor in an industrial economy (see

Handler, 1995, for an excellent review of the evolution of the ideologies behind historical iterations of welfare). However, welfare programs also served a larger purpose in a fordist regime, that of a regulatory mechanism to ensure that a potential source of labor was continuously available, to allow corporations to lay off workers periodically and keep them ready to return, and to serve as a deterrent to those who might find a life outside the confines of waged labor attractive. With the economic transformation we are currently experiencing, however, the logic at the base of the Keynesian welfare state, which in turn buoyed the fordist production regime, has melted away, to be replaced by a yet-to-be-determined postindustrial system of regulating the poor and supporting the economy.

One group of scholars identifies an acute awareness on the part of capital interests of economic restructuring and the opportunities it holds for increased power over labor:

> The central political question, then, is who will bear the hardships associated with the transition to a postindustrial economy. Will the new technology and new international division of labor lead to an improved standard of living for the average person—for example, through shortened and more flexible hours of work? Or will the final result be widespread poverty and desperation? Already, the business community has made its answer plain. It has launched a broad-scale attack on working-class standards of living, including intensified union-busting, demands for concessions in wages and benefits, and the imposition of a greater tax burden on the poor and the middle class relative to the rich. (Block, Cloward, Ehrenreich, & Piven, 1987, p. xii)

Although the details of a system comparable to the Keynesian welfare state remain cloudy, it is clear that in a globalizing economy, capital interests are likely to ensure the existence of a labor force willing to work for the low wages and low benefits, and on the contingent terms generated by the demands of accelerating international competition (Peck, 1996).

But it is politically difficult to simply do away with systems of benefits and long-standing entitlements for poor mothers and their children. The solution to this difficulty employed by the Republican Congress elected in 1994 and by an increasingly centrist Democratic president (himself caught in an election-year bind of bill-signing and promise-keeping) was to utilize the tripartite strategies of incremental erosion: *devolution, privatization,* and *dismantling.* Eliminating "big government" (i.e., dismantling the structures that buttressed a fordist regime) involved passing responsibility to lower levels of government, private providers, charitable agencies, and (perhaps most significantly) to *individuals.* The result of this is, rhetorically at least, that the states have more "flexibility" and "the people" have more direct control over how the poor are managed, while the heads of needy families are forced to "take personal responsibility" for their children's welfare. Through these processes, entitlements and guaranteed benefits are gradually eliminated.

The results of these trends and policy shifts are not yet fully known or understood, but there is a growing sense that predictions of the development of a "workfarist state" (Jessop, 1993; Peck, 1996) along Schumpeterian lines are quickly proving correct. In this scenario, capital interests are "freed" from regulatory constraints, and the poor are no longer entitled to benefits. Workfare "withdraws universal rights of access to welfare and asserts the primacy of the market as an allocative principle," and "the emblematic shift from welfare to workfare is associated with movement away from meeting social needs and toward meeting business needs" (Peck, 1996, pp. 187, 193). It is also clear that these shifts affect all social strata: "the affluent control a greater share of wealth and income than ever before; the poor are becoming both poorer and more numerous; and the middle class, faced with stagnating wages and diminishing middle-income employment, is shrinking" (Block et al., 1987, p. xii).

The geographic unevenness of the devolution, privatization, and dismantling of social welfare is chilling in its prospects. Wages and benefits are "competitively downgraded" (in Peck's terms) as localities are increasingly engaged in last-ditch attempts to attract businesses with their pliable, low-cost labor force while they simultaneously try to cope with the massive influx of unskilled young mothers who have been shunted into the labor market and who, in turn, represent a further downward force on wages. On an individual level, workers are increasingly forced to take worse jobs at lower wages under conditions beyond their control or influence. Similarly, the spatial unevenness of charitable aid, as alluded to by Wolpert (this volume), concretizes the process outlined above by Block et al. (1987) with a vicious geographic twist: the rich *areas* get richer and the poor *areas* get poorer.

These processes of devolution, as they relate to the regulation of the poor, are what I am referring to as the flexible dispersion of labor. The workfarist state, in whatever guise it may be taking, is being constructed on the basis of welfare policies that are by design flexible—they are fragmented, contingent, uncertain, reactive, and temporary—and they are organized around the goal of spatially dispersing and politically diluting labor in order to regain the upper hand for capital accumulation processes. In the flexible *accumulation* of capital, small-scale business operations and individuals are increasingly depended on to rationalize and absorb the demands and contradictions of large-scale economic forces. Similarly, in flexible *dispersion* of labor, smaller scale social welfare operations and individuals are called on to rationalize and absorb the demands and contradictions of large-scale economic forces. For localities, this means either racing to the bottom of social service provision or using their rhetorical "flexibility" to generate innovative programs that cost more than is feasible to spend. For individuals, this is the putative final incentive to "take responsibility" for their families at any social cost, any wage, and any terms defined by capital. More broadly, flexible dispersion of labor involves "attempts to resubordinate labor" (Peck, 1996, p. 232), finely tuning the labor component of postindustrial capital accumulation.

■ The Discourse of Devolution

Along with these structural shifts, and indeed, often obscuring them from the public view, has been the successive development of a political discourse centered around the themes of "small" government, individualism, responsibility, and a growing public expertise in the punitive actions that are needed to toughen up the poor. This discourse contributes to a rhetoric of devolution through a view of social justice that plays on the disgruntlement of the middle class: Justice is not being done because "handouts" are causing dependency, affirmative action constitutes favoritism, and bootstraps are not being sufficiently pulled. The myth of the meritocracy is fueled by these attitudes. Political appeals to the middle and wealthy classes include the lachrymose success stories of hard-working Americans, which, when paired with tabloid-style news coverage of welfare cheaters who somehow managed to afford a new car but kept their children in rags, foster the sentiment that individuals have been given every opportunity and still manage to depend on the state. Statistics and figures *debunking* some of the welfare myths and stereotypes ("babies having babies," generational dependency running rampant, more children being born to provide more cash, and so on) are not nearly as interesting as news bites as are media demonstrations of why welfare causes people to be lazy, cheating, and prolific breeders.

The shift toward regulating parental behavior is an important one. As of this writing, 38 states have gained approval for experimental programs that are couched in the rubric of welfare reform but that in practical terms are efforts to force parents into greater responsibility for their children through rules, requirements, sanctions, and penalties. Individual responsibility has become the catch phrase for welfare reform, which not only defuses questions of the system's shortcomings, but also lets government, at all levels, off the hook. Consistent with these more conservative messages, issues such as racial discrimination in hiring, uneven educational funding, deindustrialization, unemployment, and "spatial/social mismatch" between jobs and workers have generally fallen out of the equation (though see Nightingale & Haveman, 1995, and Peck, 1996, for analyses of labor market influences on welfare). The absence of these types of issues from the discourse also suggests the deeply aspatial assumptions made by the framers of welfare policy. Success stories of welfare-to-work programs do not come from places that face the challenges of structural decline in a globalizing economy (e.g., Detroit, Buffalo) or the desperately situated rural periphery (e.g., the Mississippi Delta region), but rather from areas that have resources to invest (e.g., Riverside, CA—see Handler, 1995) or unusually low unemployment rates, such as Wisconsin.

Nonetheless, the practical focus of state-level welfare reform programs has shifted toward a myriad of experimental projects that aim both to make the transition from welfare to work easier (e.g., child care, job training, education, "disregards" for personal assets and grandparents' incomes) and to make "noncooperative" parents suffer (cessation of benefits after specified time, no increase in benefits for subsequent children, and "full family sanctions"). As a result, we saw, prior to the passage of the 1996 welfare reform, a variegated landscape of social services and benefit levels. The idea of welfare as a service provided through national-level efforts should have been discarded much earlier, as these state- and local-level programs reshaped the geography of welfare (see Table 10.1).

So while welfare as a federal program for all practical purposes has ended, it is important to note that the focus on individual responsibility is mirrored and reinforced by rhetorical themes used in the welfare reform debate at the national level. For example, in his remarks to the *National Governors' Association* in the summer of 1995, President Clinton identified five criteria, including the following:

> First, requiring people on welfare to work and providing adequate child care to permit them to do it. . . . Second, limiting welfare to a set number of years and cutting people off if they turn down jobs. Florida got approval to limit welfare, provide a job for those who can't find one, and cut off those who refuse to work. Why not all 50? Third, requiring fathers to pay child support or go to work to pay off what they owe. . . . Taxpayers should not pay what fathers owe and can pay. Why not all 50 states? Fourth, requiring under-age mothers to live at home and stay in school. Teen motherhood should not lead to premature independence, unless the home is a destructive and dangerous environment. The baby should not bring the right and the money to leave school, stop working, set up a new household, and lengthen the period of dependence instead of shortening it. (Clinton, 1995)

Unspecified in these criteria is what happens to children when their parents are "cut off," when a parent refuses a job, when teen mothers cannot or will not live at home. Further, among the states' proposals are regulations that limit or eliminate welfare benefits to mothers who do not establish the paternity of their child, to parents who have additional children while receiving AFDC, to parents who fail to get their children immunized, and even to those who move from another state whose benefits are lower. As states are let loose on their poor families, more creative attempts to regulate adult behavior will be framed in the rhetoric of helping the children of these families; and yet both the ethics and the practical feasibility of these strategies have received little critique at the federal or state levels. Such critiques would quickly be classified as sympathetic to big government, a politically suicidal sentiment in recent years. The line between forcing parents to become part of the waged labor force and punishing their children becomes increasingly fine as "regulating the poor" (Piven & Cloward, 1971) is itself restructured.

This discourse of devolution has thus served to obscure several basic truths about welfare "reform." First, the processes of devolution, privatization, and dismantling have deep-seated gender and racial significance. Consider one example of why gender and race matter for welfare reform. Over 90% of welfare recipients are women (ACF *Fact Sheet*, 1996), which means that work requirements will be pushing poor mothers into the labor force, where, because gender-based pay inequities and

TABLE 10.1 Policy Areas and Scale for Programs Implemented Under State Waiver Program

State	Education and Training	Work	Resource	Family	Minors	Child Care	Health Care	Social Asst.	Trans. Asst.	Enforcement	Time Limits
AZ	S	S	S	S	S	S	S		S		S
AR				S	S			S		S	S
CA	S	S	S&R	S	S	S	S		S	S	S
CO	S		S				S				
CT		S	S	S	S	S	S	S	S		
DE	S	S	S&R	S	S	S	S	S	S		S
FL		S&R	S&R	S	S	S		S	S	S&R	
GA		S&RS	S		S				S	S	
HI	S&R										
IL		S	S	S	S	S		S	S	S	
IN		S	S	S	S	S		S	S	S	
IA	S	S	S	S	S	S	S		S	S	
LA				S					S	S	
MD	R	S&R	S	S			S		S	S	
MA		S	S	S	S		S	S	S	S	
MI		S	S&R	S		S			S	S	
MN	S	S	S&R		S	S	S	S	S	S	S
MS	S	S	S	S	S				S	S	
MO	S	S		S			S	S	S	S	
MT		S	S			S	S		S	S	
NE	S	S	S		S	S	S		S	S	S
NY	S&R	S&R	S		S				S	S	
NC	S	S	S	S	S		S		S	S	S
ND	S	S	S			S			S	S	

188

State	Education and Training	Work	Resource	Family	Minors	Child Care	Health Care	Social Asst.	Trans. Asst.	Enforcement	Time Limits
OH	S		S	S	S	S	S	S	S	S	S
OK		S	S		S			S		S	
OR	S	S&R	S	S	S	S	S	S	S	S	
PA		S&R			S	S		S&R	S		
SC	S	S&R	S&R	S	S	S	S	S	S	S	S
SD		S	S				S		S	S	
TX		S	S			S	S		S	S	S
UT		S	S&R		S	S		S&R	S		
VT	S	S	S	S	S	S	S		S	S	S
VA		S								S	
WA		S	S						S		
WV		S	S	S			S	S	S	S	S
WI	S	S	S		S					S	S
WY		S			S					S	S

1. Policy areas have been coded to 11 categories in order to simplify the presentation of material. Note, for instance, that most states implementing programs focus on work, family resources, enforcement of program rules, and time limits. Some examples follow to clarify the meaning of the less obvious category names: "Resource," any specifications regarding resources or assets of the household; "Family," any family-oriented rules regarding who the members of the household are; "Minors," any specifications regarding rules for minor parents; "Health Care," any transitional medical insurance as well as child support, paternity establishment; "Social Assistance," any programs aiding adjustment of the parents or household, including parenting classes, legal services, help in locating treatment for addictions; "Transitional Assistance," any type of short-term assistance for newly employed parents (e.g., child care, health care, transportation).

2. S indicates a state-level program or policy, S&R indicates both state- and regional-level, and R indicates regional-level only.

3. This table includes data available as of July 1996 from the Administration of Children and Families of the U.S. Department of Health and Human Services for the 38 states that have applied for waivers. The states that had not, at that time, applied for waivers are AL, AK, ID, KS, KY, ME, NV, NH, NJ, NM, RI, and TN.

gender divisions of labor still exert strong downward pressure on women's average wages, their full-time earnings will average three-quarters of men's (see Table 10.2). If we consider too that approximately two thirds of welfare recipients are Hispanic or African-American, the effects of racial discrimination force average wages for women to just over half those of white men. And when we consider that, because of the contingent terms of work at the low end of the labor market, many of these women will be working part-time in short-term jobs, their potential earnings shrink further still. Finally, a plurality of welfare recipients fall in the lower categories of educational attainment, an additional downward force on average wages. Wage differences such as these have not been accounted for in any of the demands for personal responsibility through waged labor.

Second, the discourse of devolution masks the glaringly obvious fact that the problems of poverty are fundamental to the existence of a welfare population (Handler, 1995). Discussions of "workfare," time limits, block grants to states, and "illegitimate" births rarely mention poverty because there is a preoccupation with individual responsibility and hard work. The underlying message is that the causes of poverty lie with the *individual*—not with economic shifts, exploitation, race or gender discrimination, disinvestment in education and social supports, or a lack of available jobs. Political and capital concerns are to end *welfare,* not to end *poverty.*

Third, it is rare too for the political rhetoricians to recognize that there are spatial variations in labor markets and the characteristics of needy populations, and that these merit an approach sensitive to regional and local contexts. The discourse of devolution obfuscates the fact that there are insufficient jobs available for current welfare recipients; that those that are available are most often part-time, short-term, and low-waged, a combination not conducive to supporting a family; and that the majority of welfare recipients live in either rural or central city areas, while most low-skill jobs are in the suburban service sector (see Handler, 1995, for detailed examples). Further, these jobs very rarely carry with them health insurance and other benefits, nor do they generally make provisions for child care.

Fourth, two unheralded roles of the existence of the very poor and the sanctions they experience are that of social "deterrent" for the near-poor and marginally employed, and that of an economic bottom-rung that serves to depress the entire wage structure of the lower end of

TABLE 10.2 Median Weekly Earnings of Full-Time Workers, by Sex, Ethnicity, and Education, 1996

Race/Ethnicity:	Men's earnings	Women's earnings	Women's % of men's	Women's % of white men's
White	580	427	73.6	73.6
Black	417	372	89.2	64.1
Hispanic	344	315	91.6	54.3
Education:	$	$	%	% Women's earnings are of Men's Avg.
Avg. of 25-yr.-olds	597	442	74.0	74.0
Less than High School	343	268	78.1	44.9
H.S. Diploma/GED	509	363	71.3	60.8
Some College	600	433	72.1	72.5

Source: *Bureau of Labor Statistics,* Report on Median Weekly Earnings for the First Quarter, 1996.

the labor market. These roles too have geographically uneven expressions, resulting in what Peck (1996) calls "workhouse regions" (p. 253), which represent a continued bifurcation in both personal and regional wealth and poverty.

Finally, the discourse of devolution is at least partially responsible for the political primacy of focus on welfare reform to the near exclusion of other anti-poverty programs that have garnered some significant success. For example, expansion of the Earned Income Tax Credit has not been made a priority despite its track record of putting more money into the hands of the working poor who need it most. The minimum hourly wage has been raised, but it is still insufficient to bring a single parent and her child above the poverty line in a full-time job. The desperate need for health insurance and less expensive medical care for working families has still not been adequately addressed, despite recent "portability" legislation. And the continuing dilemma many families face of finding quality child care at reasonable cost is still virtually ignored by political elites (see Handler, 1995). Increasingly, then, the working poor are becoming more vulnerable at the same time that welfare recipients are being forced into the low end of the labor market, creating the conditions for a potentially disastrous collision.

■ Recent Trends in the Devolution of Welfare

Recent alterations in social policy reflect these discursive developments. In the period immediately following the election of the Republican Congress in 1994, the Clinton administration and Congress have both promoted various solutions to the welfare "problem." With the expanded state waiver program, time limits, work requirements, school attendance and residence restrictions for teen mothers, and sanctions for noncooperative recipients of AFDC have already gone into effect in many areas (Table 10.1), creating a spatially differentiated landscape of social welfare provision. Through these programs, the shift of responsibility for administration of welfare away from the federal government and toward states and localities had already begun before the PRWOR was passed in August of 1996. Simultaneously, a renewed shift of responsibility for the financial support of poor children toward individual parents by requiring *some* type of work dominated political campaign speeches and was reflected in the names of state-level programs.[2] Indeed, during this period there was a growing sense among the public that if only the federal government stopped interfering, states would be more innovative with programs and individual parents would be motivated to get off welfare rolls and into jobs.

In these popular sentiments, and, more concretely, in the policies that are already being implemented, we can identify general strategies of state restructuring that are reflected in the specific restructuring of the U.S. welfare system. In this section, I first identify recent developments in national welfare policy to familiarize the reader with the very complex system of public assistance and to set the context within which restructuring is taking place. In the following section, I identify ways in which state restructuring is changing the *scale* and *scope* of government with regard to welfare policy via strategies of *devolution, privatization,* and *dismantling* (Kodras, this volume), demonstrating the links between the ideological justifications for recent policy changes and the implications these have for state restructuring.

Excellent reviews and analyses of the development of social policy in the United States exist (e.g., DiNitto & Dye, 1987; Piven & Cloward, 1971; Skocpol, 1992), and thus I do not intend to provide a great deal of historical context here. Rather, I will consider the more recent developments that are linked to the current round of economic restructuring, with Aid to Families with Dependent Children (AFDC), the program most commonly referred to when the term "welfare" is used in the United States, as the focus.

AFDC provides cash income to the parents of children under 18 years old who, for various reasons, have no (or limited) sources of income and minimal assets. Beginning as "widow's pensions" to support fatherless families (Skocpol, 1992), AFDC underwent rapid changes and expansion in the late 1960s and 1970s as more women entered the waged workforce, divorce rates and single motherhood rose rapidly, well-paying jobs for unskilled and semi-skilled industrial workers declined, and wage disparities grew. In the late 1970s, however, the beginnings of the retrenchment in welfare can be identified, and the 1980s saw a steady deepening of this retraction. As the competition for capital accumulation rapidly accelerated and became globally oriented, the erosion of entitlements and the expansion of demands for individual accountability began in earnest with the Family Support Act of 1988. These processes of retrenchment of the welfare state have continued into the mid-1990s to the degree that federal guarantees of support have now been dismantled, work requirements are in place, and the discourse of individual responsibility is dominant.

As of this writing, the PRWOR has not yet taken effect; therefore, the current situation, although imminently changing, is discussed here. In terms of regulatory and economic structure, AFDC has involved a partnership between the federal and state governments:

> Federal and state governments share in its cost. The federal government provides broad guidelines and program requirements, and states are responsible for program formulation, benefit determinations, and administration. Eligibility for benefits is based on the state's standard of need as well as the income and resources available to the recipient.... The federal government reimburses the states for operating an AFDC program with matching funds ... [which] may range from 50 percent for states with the highest per capita income to 83 percent for the state with the lowest per capita income. (ACF *Fact Sheet,* 1996)

Thus, cash benefits from AFDC vary significantly throughout the country; the Northeast, Wisconsin, and California consistently have the highest rates, whereas the Deep South typically has the lowest (Kodras, Jones, & Falconer, 1994). Along with AFDC, many families also qualify for Food Stamps (see Kodras, 1990) and school lunches, housing subsidies such as Section 8 grants, Medicaid health coverage, Head Start early childhood education, and a range of locally provided assistance in the areas of legal advice, transportation, supplementary education, low-interest business start-up loans, and other services. Eligibility for AFDC has been limited to low-income households with one head and at least one child under 18. The majority of recipient households are headed by single women; however, a much smaller program, AFDC-UP, allows two parents in the household and provides linked AFDC and unemployment benefits, along with job assistance, for families in which the principal earner is unemployed.

The formal addition of work requirements for welfare recipients is, as suggested in the previous sections, one of the most significant developments in welfare policy in the past decade. State-level programs implemented under the waiver process combined work requirements with time limits (generally around 2 years), which have been proposed as the "stick" to provide an added incentive for joining the waged labor force. Both of these are found in the 1996 bill. At the root of these developments is the vocal reestablishment of the necessity for welfare to be a "second chance, not a way of life," that is, for welfare to be a family's transitional helping hand rather than a permanent minimum income guarantee.

The idea of "workfare" is not new, having roots in the "workhouses" of the 19th century and earlier (Handler, 1995; Skocpol, 1992), and more recently in the Works Progress Administration of the New Deal and various post-World War II efforts to provide a minimum safety net. The Nixon administration's attempts at a Family Assistance Program and Carter's "Program for Better Jobs and Income" were only marginally implemented and were mortally wounded by Reagan's Omnibus Budget Reconciliation Act in 1981, which slashed social programs and aid to the poor (Glazer, 1995). However, on the eve of Reagan's departure in 1988, the Family Support Act (FSA) was framed and passed by the Democratic Congress. The FSA formalized two basic shifts in welfare policy that are central to today's debate: first, it put renewed emphasis on moving recipients into training and work, largely through the establishment of the Job Opportunities and Basic Skills (JOBS) Training Program, and second, the FSA gave greater latitude to individual states to experiment with alternative methods of moving people into the waged workforce (for comprehensive reviews of FSA and JOBS, see Gueron & Pauly, 1991; Manski & Garfinkel, 1992; Nightingale & Haveman, 1995). The Administration for Children and Families (ACF), a branch of the federal Department of Health and Human Services (DHHS), spells out the changes these brought:

> Passage of the Family Support Act and the establishment of JOBS reflect a rethinking of the welfare system. It no longer merely provides cash assistance to meet the basic needs, but now encourages economically disadvantaged people and families to gain skills that allow them to move permanently into the economic mainstream, while cash assistance is considered transitional. (ACF *Fact Sheet*, 1996)

These two shifts can be seen as strategies of devolution: responsibility is imposed on the individual and the states as the federal government removes itself from the arena of direct support for the poor.

Although states have long played central roles in the administration of programs such as AFDC, the Family Support Act promoted greater flexibility for states, in part through the use of legislative "waivers" that allow states and even local areas to experiment with different strategies. The Bush administration granted a small number of these waivers, but the Clinton administration has been far more aggressive in this practice, granting an unprecedented number of waivers covering two thirds of the nation (ACF *Fact Sheet,* 1996). President Clinton's speech to the National Governor's Association in the summer of 1995 promised the state governors that if they submitted proposals that target at least one of five identified criteria for welfare reform, they would receive approval for their projects within 30 days (Clinton, 1995). In a strategic move, Clinton asked the governors to support the latest welfare reform bill, but simultaneously opened the way for states to circumvent Congress, therefore both speeding up the "reform" process and granting individual states greater power in determining their welfare provision policies. The President said:

> [W]e don't have to wait for a Congress to go a long way toward ending welfare as we know it, we can build on what we've already done. . . . Already in the last 2-1/2 years, your administration has approved waivers for 29 states to reform welfare your way. The first experiment we approved was for Governor Dean, to make it clear that welfare in Vermont will become a second chance, not a way of life. Governor Thompson's aggressive efforts in Wisconsin, which have been widely noted, send the same strong message. Now, we can and we should do more, and we shouldn't just wait around for the congressional process to work its way

through. We can do more based on what states already know will work to promote work and to protect children. (Clinton, 1995)

Despite its relatively small piece of the budget and small population served, the restructuring of welfare has been a key component of the Clinton administration's efforts to demonstrate the end of "big government" and increased local power or, from the perspective developed in the previous sections, the end of the Keynesian welfare state and the devolution of welfare to a scale and design more in line with the flexible capital accumulation tactics of post-industrial capitalism that can take advantage of the geographic unevenness in the social regulation of labor—that is, the flexible dispersion of labor.

■ Changing the Scale and Scope of Welfare Policy

Kodras (this volume) notes that three strategies can be identified as common elements of state restructuring. First, *devolution* involves the displacement of an operation to a lower tier of government. Second, *privatization* occurs when an operation is moved from the "public" arena of government into the "private" arena of the market and the non-profit sector. Third, *dismantling* is the term used for the cessation of a particular operation or set of operations; that is, it is deemed that the function no longer needs to be performed by either the government or the private sector. There is evidence of all three in recent welfare "reforms."

Devolution of Welfare

The federal government has already begun the process of "devolving" the operations of social welfare to lower tiers of government but the question of the desirability of these moves is still open (witness the dissatisfaction many Democratic members of Congress felt at the time of President Clinton's signing of the reform bill). The justifications— again, symbolic of the discourse of devolution—behind "ending welfare as we know it" by shifting administrative and fiscal responsibilities to the state and local levels are to minimize "big government" and the magnitude of federal control over public assistance. Passing welfare down to the individual states involves a *scale* change as the national government is gradually removed from the business of providing for poor families and the states are given responsibility for their most economically and socially powerless residents and for crafting regulatory social policies. Paralleling the Federal Agricultural Improvement and Reform Act of 1996 (FAIR) (see Page, this volume), which eliminated many long-standing farm subsidies, this move could be seen as the states' "Freedom to experiment with poor families" initiative, with the hope that states will succeed where the nation failed to develop ways of regulating poverty, without the assistance/interference of the federal government. The rhetoric continues through the assumption that by giving states the "freedom" (as well as the responsibilities) of public assistance, new and innovative methods of providing transitional help to needy families will be developed in ways that the federal government could never have tried at the much larger scale of the nation. After a period of innovation, the best methods and programs will presumably be copied by other states in translocal adoption patterns and the end result will be a leaner, tougher system that works as well as possible at getting people out of dependency and into the economic mainstream.

There is also a change in *scope* here. The federal government's position as the establisher of minimum levels of protection, broad policies, and widespread mandates has already started to shrink and may be eliminated entirely. Smaller units of government are seen as more efficient, more responsive to local needs, and better able to change quickly in response to fluctuating economic, political, and social conditions. Some areas of concern will become more important than others, depending on local context. Thus in one part of the country the scope of welfare may include a heavy emphasis on furthering education, whereas in another area greater weight may be placed on maximizing collection of child support from noncustodial parents. One state may focus primarily on work incentives for parents through training and application assistance, whereas another concentrates on early childhood health and education through immunization requirements and mandatory parenting classes. The scope of possibilities through which state-level welfare agencies can delve into and regulate the lives of poor families is more open than the federal government's recent array of programs due to the greater flexibility of the states. As a result, we can expect to see further geographic differentiation in social welfare provision.

There is, however, resistance to allowing social welfare policy to devolve to states, stemming both from sentiments that the federal government should play a role in welfare policy and from worries that states cannot or will not provide adequate services. A year before signing the PRWOR, President Clinton himself pointed out the danger of too much state power and flexibility, drawing on his experience as the governor of Arkansas and rejecting then-Senator Dole's idea of block grants as he spoke of welfare to the country's governors:

> I also have to tell you that I am opposed to welfare reform that is really just a mask for congressional budget cutting, which would send you a check with no incentives or requirements on states to maintain your own funding support for poor children in child care and work. And I do believe honestly that *there is a danger that some states will get involved in a race to the bottom,* but not, as some have implied, because I don't have confidence in you, not because I think you want to do that, not because I think you would do it in any way if you could avoid it, but because I have been a governor for 12 years in all different kinds of times. And I know what kind of decisions you are about to face if the range of alternatives I see coming toward you develop. . . . I don't know what your experience is, but my experience is that the poor children's lobby is a poor match for most of those forces in most state legislatures in the country. Not because anybody wants to do the wrong thing but because those people are deserving too and they will have a very strong case to make. (Clinton, 1995; emphasis added)

The "race to the bottom" problem has been noted in other policy areas in which local or state units attempt to minimize expenditures, particularly through short-term, piecemeal methods of squeezing every dollar in tough economic times (Peterson, 1995). Thus, the ponderous and highly contested path of federal legislative change may in some cases be preferable to the potential of quick, razor-sharp slashes by state governments. Further, in the area of program evaluation, the long-term effects of policy changes are not immediately known and cannot be accurately assessed without fairly extensive (and often expensive) periodic reviews, which states may be less likely to fund than the federal government has been (Gueron & Pauly, 1991; Manski & Garfinkel, 1992).

A second area of resistance to the devolution of welfare policy emerges from concerns surrounding the geographically uneven distribution of benefits and programs. In part because of spatial variations in incomes and cost of living across the country, AFDC benefits are already geographically uneven. Jones (1990), however, found that these disparities are also affected by racial discrimination, unemployment, and uneven job markets throughout the country. Further, specific *places* have differing traditions and commitments based on political practices, economic characteristics of the region, and social relations, such as racial conflicts (Kodras, this volume). Given such unevenness within a nationally regulated welfare framework, disparities are likely to be greatly exacerbated as states are granted unprecedented levels of discretion in the realm of welfare policy.

Finally, devolution of social welfare administration could be critiqued from the point of view that the large-scale and long-term pooling of both finances and risks minimizes the negative impacts of contingency. States (particularly poorer ones) are less able to fund projects that the federal government could afford through averaging costs among many units or amortizing expenditures over long periods of time. Further, if block grants are implemented, there is the added question of growing competition between states and between localities within states for scarce resources. As Kodras (1990) has shown with the Food Stamp program, the most *organized* and politically experienced, rather than the most *needy,* areas tend to be most successful in obtaining funds. Thus, although subnational governments may be more flexible and able to innovate, they also are more vulnerable to fiscal crises, program failures, and competition.

As of this writing, the White House is steering a course between the two extremes. The federal guarantee of support has been technically withdrawn, and states are being granted greater latitude in experimenting with welfare policies. But they are (at least for now) doing so under the supervision and guidelines of the Department of Health and Human Services. There is a sense that states should be given a chance to innovate; but despite the passage of the Welfare Reform Bill, there remains a faction that believes that there should still be a nationally regulated and guaranteed safety net for poor families and their children. The question, then, is no longer whether or not the devolution of welfare is occurring, but rather, what form these processes will take.

Privatization of Welfare

Should the operations of social welfare be shifted away from government and toward the free market and the voluntary sector? To a significant degree, this shift is already occurring. The final point of President Clinton's five criteria for state programs (above) was to involve businesses in the welfare-to-work effort by allowing states to pay the cash value of welfare and food stamps to private employers as wage subsidies. He said:

> This so-called privatizing of welfare reform helps businesses to create jobs, saves taxpayers money, moves people from welfare to work, and recognizes that in the real world of this deficit we're not going to be able to have a lot of public service jobs [to give] to people who can't go to work when their time limits run out. (Clinton, 1995)

The implications of this approach are clear: The era of government-based *public* entitlement to assistance is seemingly drawing to a close, but whether the true beneficiaries will be poor children or businesses remains to be seen. Businesses are motivated to become involved in welfare-to-work programs, spurred in part by developing programs that pay the wages of new workers for a set amount of time (as in President Clinton's remarks, above) in exchange for free workers and then trained, low-wage workers. Is it only a matter of time before private consulting groups are contracted to manage local welfare-to-work programs, as they have already been hired to manage local school districts? Further, the nonprofit sector (charities, churches, community groups, etc.) is already struggling to meet the growing needs of an increasingly bifurcated society (Wolpert, this volume).

The rhetorical justification for at least partially privatizing social welfare provision is again the "end of big government" argument, whereby shrinking the scope and scale of the federal government is seen as desirable in order to minimize interference by an unwieldy system. Justifications also lie in the conservative argument that the free market can act as an "allocative principle" (Peck, 1996), and the rest can be left to the volunteers.

But at a more fundamental level, the issue of privatizing welfare raises the question and challenges the notion of what constitutes the public arena and what responsibilities lie there. Classical social contract theorists like Hobbes and Rousseau suggested that the negotiations between individuals and society necessarily involve some give and take, some exchanges of freedoms and assurances, and tradeoffs between privacy and certain societal guarantees. Modern writers have also pointed out that not everyone is equal in the contract and that race-, class-, and gender-based social relations skew the public and private spheres in favor of some social groups over others (Pateman, 1988). The current debate over welfare includes many discussions of rights, opportunities, responsibilities, and requirements, which indicate a deeper level of concern for governance beyond the details of time limits and job training. Thus, privatization is more than just involving nonpublic sectors in getting people off welfare; it is also about democracy, citizenship, and accountability.

There is, of course, an argument against further privatization. First, the free market has no immediate interest in providing livelihoods to poor families except in extreme cases in which hungry, pliable, and locally dependent workforces are needed (Block et al., 1987; Piven & Cloward, 1971). Second, the nonprofit sector cannot absorb the additional burden of supporting 5 million American families at a time when volunteer hours are decreasing, the middle class is working longer hours for lower wages, and the sense of community is reported to be under siege (Putnam, 1995). Further, even if local communities did look after their own (harking back to the pre-New Deal methods of "indoor" poor relief), the vast spatial differentiation of wealth in local areas means that the poorest areas with the greatest need would have the fewest resources to address that need (Wolpert, this volume).

At the present time, the federal government is setting a middle course that combines some degree of financial backing (through block grants) and administrative guidance (through state eligibility requirements) of the federal and state governments with private businesses' training and workfare programs, and combines public assistance with nonprofit programs. The result is a crazy-quilt of programs and types of assistance, ranging in both scale and scope from basic AFDC regulations to pro-bono legal assistance and church-run soup kitchens in local areas.

Dismantling Welfare

Should the operations of welfare even continue to exist? People who lose out when programs are dismantled or scaled back do not just disappear. When the Reagan administration deinstitutionalized hundreds of thousands of mental patients, some were taken in by their families, who absorbed the costs and difficulties, whereas many others became homeless, were convicted of various crimes and sent to prison, or died. Similarly, if Medicare for the elderly is cut back severely, either families will struggle to care for their elders or individuals will be left to fend for themselves. There is no reason to think that the dismantling of welfare would result differently, and yet there are currently both discussions of this possibility and concrete proposals aimed in this direction.

The rhetorical justifications for dismantling welfare again come back to arguments *against* big government and *for* individual responsibility and hard work. A shift toward individual responsibility, work not welfare, and "get-tough" political stances on recipients has dominated the discourse in welfare debates of the mid-1990s. In this line of thinking, parents are accountable and should be required to work at whatever job they can, practice frugal household spending measures, and instill in their children a solid work ethic. For example, in speaking of providing quality child care, President Clinton said, "This is a very big issue if your objective for welfare reform is *independence, work, good parenting, and successful children*" (Clinton, 1995; emphasis added). Similarly, the DHHS's Administration for Children and Families (quoted above) attempts to convey the values of self-sufficiency, entering the economic mainstream, and work. The transitional nature of public assistance also indicates an expectation that such government programs will someday not be needed for any "able-bodied" adults, an altogether unrealistic assumption, given the vagaries of job opportunities as places shift position within the global economy.

Despite President Reagan's 1981 budget slashes and subsequent questions about the very existence of social welfare policies and public assistance that we have experienced in the 15 years since then, there seems to remain a sentiment among many politicians and much of the public that some sort of transitional "hand up" should be available for needy families. However, this sentiment is frequently reserved for the so-called "deserving" poor, who are disadvantaged by situations not of their own making (see Handler, 1995, for a discussion of past conceptualizations of "deserving" and "undeserving" poor). But two serious concerns arise from such a distinction. First, these labels are created by political, economic, and social elites. The successful labeling of particular groups as "undeserving" allows them to be discarded, and another group attacked. These processes of labeling enable the continued erosion of entitlements as social welfare is steadily dismantled piece by piece.

The second concern arises from how we define a situation of one's own making in an era in which the global economy is changing so quickly that places, people, and policies cannot keep pace. Consider again the above argument that the current economic transformation is due to rapid shifts in international capital accumulation and concomitant labor dispersion, and that the deep restructuring of production and retrenchment of wages, benefits, and entitlements are playing a major role in squeezing more and more people out of a fordist wage economy into fragmented, contingent, and uncertain work conditions, or into long-term joblessness. If the "deserving poor" are defined as those whose situations are not of their own making, how can we determine who is deserving when public and political discourse ignores the structural causes of poverty?

We are, in fact, seeing some evidence of the dismantling of welfare among those for whom the label "deserving" has been steadily eroded. Unmarried teenage mothers are increasingly required by states to stay in school and live at home, despite the fact that these requirements do not apply to nonparents over 16 years, and are not, in fact, required of unmarried teen *fathers*. Similarly, state programs are increasingly mandating the establishment of the child's paternity in order to collect benefits. In addition, there are requirements for child immunization, job training, cessation of benefits for "noncooperative" parents (i.e., those who refuse a job or do not attend training), and continued shrinking of time limits. Although all of these might be interpreted as helping people move off welfare rather than as indicative of the dismantling of social policy, it is important to note that they are all focused on holding the individual and the family accountable for their situation, regardless of the greater structural causes of poverty and joblessness. It is this shift toward a solely *personal* responsibility, to the exclusion of structural or contextual factors, that portends a trend toward dismantling.

■ Conclusions

Welfare is actually a very small piece of the federal budget and serves a relatively small proportion of the population, yet it receives a disproportionate amount of attention. Due to the political weakness of its beneficiaries, welfare is more likely to be shaped, formed, changed, devolved, privatized, and dismantled in ways that Medicare, Social Security, the military, and so-called corporate welfare programs will never be. State restructuring is a rapidly evolving and constantly changing phenomenon, and yet it is possible to identify common strategies. Welfare policies have been subject to these strategies of devolution, privatization, and dismantling during the past decade. These policy changes must be placed in a larger context that identifies both structural transformations and discursive shifts, as well as the way that these affect the redesign and reduction of the welfare system through the erosion of entitlements and benefits.

More broadly, the welfare state has seemingly lost its relevance as a support for a regime of industrial capitalism that no longer dominates the U.S. economy. The assault on welfare needs to be seen as part of both the transformation brought on through economic restructuring and the increasing efforts to harness the flexible accumulation of capital with the flexible dispersion of labor. In response to these structural changes, the scale and the scope of governance are shifting to accommodate the need to shift responsibility away from the federal level to states, localities, and, ultimately, to individuals. Responsibility for, and regulation of, the poor are being devolved while at the same time retrenchment occurs in the areas of rights, entitlements, and benefits, combining to "end welfare as we know it"—but to what end remains to be seen.

NOTES

1. The author gratefully acknowledges financial support for this project from the National Center for Geographic Information and Analysis (NCGIA) and NSF grant SBR-8810917.

2. Including Arizona's "Employing and Moving People Off Welfare and Encouraging Responsibility" (EMPOWER); Colorado's "Personal Responsibility Project"; Georgia's "Personal Accountability and Responsibility Project"; Illinois' "Work and Responsibility Demonstration"; Louisiana's "Individual Responsibility Project"; Missouri's "Families Mutual Responsibility Plan"; and South Carolina's "Self-Sufficiency and Parental Responsibility Program."

REFERENCES

ACF (Administration for Children and Family) *Fact Sheet.* (1996). Washington, DC: Department of Health and Human Services.

Block, F., Cloward, R., Ehrenreich, B., & Piven, F. F. (1987). *The mean season: The attack on the welfare state.* New York: Random House.

Clinton, B. (1995, July 31). Remarks to the National Governors' Association. Burlington, VT.

DiNitto, D., & Dye, T. (1987). *Social welfare: Politics and public policy.* Englewood Cliffs, NJ: Prentice Hall.

Gans, H. (1995). *The war against the poor: The underclass and antipoverty policy.* New York: Basic Books.

Glazer, N. (1995). Making work work: Welfare reform in the 1990s. In D. Nightingale & R. Haveman (Eds.), *The work alternative: Welfare reform and the realities of the job market* (pp. 17-32). Washington, DC: Urban Institute Press.

Gueron, J., & Pauly, E. (1991). *From welfare to work.* New York: Russell Sage Foundation.

Handler, J. (1995). *The poverty of welfare reform.* New Haven, CT: Yale University Press.

Jessop, B. (1993). Towards a Schumpeterian workfare state? Preliminary remarks on post-Fordist political economy. *Studies in Political Economy, 40,* 7-39.

Jones, J. P. III (1990). Work, welfare, and poverty among black female-headed families. In J. Kodras & J. P. Jones (Eds.). *Geographic dimensions of United States social policy* (pp. 200-217). London: Edward Arnold.

Kodras, J. (1990) Economic restructuring. Shifting public attitudes and program revision: The politics underlying geographic disparities in the Food Stamp program. In J. Kodras & J. P. Jones (Eds.). *Geographic dimensions of United States social policy* (pp. 218-236). London: Edward Arnold.

Kodras, J., Jones., J. P. III, & Falconer, K. (1994). Contextualizing Welfare's Work Disincentive: The Case of Female-Headed Family Poverty. *Geographical Analysis, 26,* 3-18.

Manski, C., Garfinkel, I. (Eds.). (1992). *Evaluating welfare and training programs.* Cambridge, MA: Harvard University Press.

Nightingale, D., & Haveman, R. (Eds.). (1995). *The work alternative: Welfare reform and the realities of the job market.* Washington, DC: Urban Institute Press.

Pateman, C. (1988). *The sexual contract.* Stanford, CA: Stanford University Press.

Peck, J. (1996). *Work—place: The social regulation of labor markets.* New York: Guilford.

Peterson, P. (1995). *The price of federalism.* Washington, DC: Brookings Institution.

Piven, F. F., & Cloward, R. (1971). *Regulating the poor: The functions of public welfare.* New York: Pantheon.

Putnam, R. (1995). Bowling alone: America's declining social capital. *Journal of Democracy, 6,* 65-78.

Skocpol, T. (1992). *Protecting soldiers and mothers: The political origins of social policy in the United States.* Cambridge, MA: Belknap.

11
Transnationalism, Nationalism, and International Migration: The Changing Role and Relevance of the State[1]

RICHARD WRIGHT

At the time of writing, it appears that U.S. immigration law concerned with numbers, skill levels, "family reunification," and the like, will for all practical purposes remain unchanged in the near future. In contrast, the federal government, from the national to local levels, is actively addressing illegal immigration by authorizing further investment in the Immigration and Naturalization Service (INS) and the United States Border Patrol, establishing pilot programs to enable employers to verify the status of newly hired workers and bolstering incarceration and deportation procedures, among other things. While Congress and other governmental units attempt to shore up U.S. borders against illegal crossers, however, the United States is increasingly involved in the transnational circulation of capital, goods, information, and labor. This chapter explores the timing of these contradictory trends and the significance of what Rouse (1995) has called the emergence of a new transnational/national dialectic.

The question of why certain types of nationalist sentiment and exclusionary policies intensify in an era of accelerating transnationalism becomes clearer when examined through the lens of international migration (Higham, 1988). Specifically, the recent ascendance of a certain type of nationalism, in the form of jingoistic oratory and exclusionary practices, results in part from threats imposed by transnationalism. As the United States becomes increasingly enmeshed within the global economy, resulting transformations in the U.S. job market—plants moving offshore, jobs becoming increasingly unstable and contingent, and wage/benefit packages declining in line with global competition—threaten specific segments of the U.S. workforce and frighten a much broader portion of the population. Immigration is seen to be part of the problem as it brings foreign workers into the increasingly competitive U.S. labor market. It is important to note that immigration is one of the few issues that the state can effectively control on behalf of groups feeling the threat of globalization.

As argued in chapters throughout this volume, the state plays a complex and often contradictory role in the present era of global restructuring, seeking both accumulation for capital and legitimacy from civil society. Put differently, the state champions both the forces of transnationalism and of nationalism. On the one hand, the state supports the global accumulation of capital through neoliberal, free-trade policies such as GATT and NAFTA. Promoting the forces of transnationalism is important because the state requires a growing economy as a financial basis for its own sustenance. On the other hand, the state must act to ensure its own legitimacy in the eyes of the electorate if it is to remain in power within a liberal democracy. As resistance to globalization grows within the electorate, the state has an increasing interest in supporting the forces of nationalism, expressed, for example, in exclusionary immigration policies that appear to lessen domestic job competition or in welfare reform that denies benefits to noncitizens. To begin to explore these contradictory roles in more depth, this chapter starts with a review of recent actions taken by the state in halting illegal international migration in the context of accelerating transnationalism and rising nationalism.

Simply twinning the "transnational" with the "national," however, provides a relatively superficial interpretation of recent events. I argue that the issue is more than dialectical, more complex than a straightforward confrontation between state-based forms of U.S. nationalism and transnationalized capital and labor. To make my case, I draw on several key perspectives that add dimension to the issue. For example, Rouse (1995) asserts that we must look beyond the institutions of the state to understand national affiliation and identity, including the mass media, churches, schools, and other institutions. Ong (1995) confronts a number of nationalist myths operating in the United States, showing, for example, that the movement of Asians (literally) toward their "American Dream" challenges the fundamental idea that the United States is a nation rooted in Anglo-Saxon cultures. In the latter part of the chapter, I use viewpoints such as these to add dimensionality to the transnational/national dialectic, demonstrating new complexities in the expression of nationalistic rhetoric and transnational migration. In so doing, I speak to the larger issue of the historically evolving relationship between transformations in the economy and the changing relevance and role of territorial states (Agnew, 1994).

■ U.S. Nationalism and Transnational Migration: The Changing Role of the State

A distinct increase in state efforts to halt illegal immigration is evident at all levels of the federal hierarchy in the past few years. At the national level, the President and Congress are presently in the process of transforming the Border Patrol through massive increases in its budget and workforce. For example, legislation pending in a House-Senate conference committee would provide for stronger border enforcement, adding 1,000 Border Patrol agents per year for 5 years, bringing the total from 5,175 in 1996 to almost 10,000 by the year 2000. If enacted, the legislation also would fund a 14-mile "triple fence" on the U.S.-Mexican border south of San Diego and would increase the penalties for smuggling aliens into the United States. In terms of state restructuring, no government agency in the past two years has been more restructured than the U.S. Border Patrol!

Efforts to curtail illegal immigration are also evident in the rhetoric and national political platforms of the 1996 presidential election. Most notably, Patrick Buchanan's serious run (and California Governor Pete Wilson's less significant run) for the Republican Party's presidential nomination was based, in part, on an anti-immigration platform that advocated closing the border to undocumented migrants, stopping entry "(C)old, period, paragraph" (Buchanan, 1996).

Clinton's restructuring of the INS and Buchanan's rhetoric link with an anti-immigrant sentiment surfacing in policies at the state level. The most notorious example is California's 1994 Proposition 187 ballot initiative, crafted to "Save Our State" and deprive undocumented immigrants of their rights to public schooling, nonemergency health care, and welfare. Another example is the "Florida Initiative," announced in May 1996 by Governor Chiles that will, among other provisions, triple bed capacity at the Krome Service Processing Center (where many illegal immigrants entering Miami are detained), establish an immigration court in Miami International Airport, enable the deportation of criminal aliens directly from jails and prisons, and launch an anti-alien smuggling operation in the Caribbean Basin.

A number of programs linking national and local forces have been initiated in recent years as well. For example, legislation currently pending in the U.S. Congress allows the Attorney General to permit local police departments to "seek, apprehend, and detain" illegal aliens who are subject to an order of deportation. In addition, the INS has established Operation Hold the Line (implemented in September 1993 in El Paso, Texas, to reduce illegal crossing), Operation Gatekeeper (implemented in October 1994 on the San Diego-Tijuana border to reduce illegal crossing in western San Diego County), and numerous other programs. Furthermore, INS officers are delegated to local jails. In Anaheim, California, for example, the INS stationed agents at the city jail to conduct a 90-day study that examines the citizenship of all inmates. Agents also screen the citizenship of criminal suspects going to court. According to the *Los Angeles Times,* Anaheim officials believe that interviewing arrestees before their arraignment is crucial. It makes a judge aware of an individual's immigration status so that a "hold" can be placed on him or her to prevent release before trial.

These various efforts by the state to strengthen the U.S. borders occur at a time when those same borders are weakened by increasingly transnational interaction and integration. The U.S. economy is peppered not only with labor but also with capital, goods, and information from many other countries (see Lake and O'Loughlin in this volume). Similarly, businesses headquartered in the United States use the opportunity that globalization presents to invest in productive capacity offshore in places like Bangkok, Tijuana, Sao Paolo, and Jakarta. The process is especially pronounced along the international border with Mexico, which has all but disappeared in economic terms. Mexico is an important origin country for both legal and illegal immigrants to the United States. Over 25% of all the foreign born who declared they "came to stay" in the United States in the 1990 census were born in Mexico, compared with the Philippines, the next most important country of origin, with only 6% of the total (U.S. Bureau of the Census, 1993). In addition, Mexicans are overrepresented among the undocumented U.S. population (Bratsberg, 1995). Mexico, of course, is also an important destination for U.S. investment, with NAFTA the notable example here.

In the United States, the internationalization of productive activity has wrought significant social and economic change. Deep economic restructuring in the United States has eliminated jobs in some sectors and created new jobs in both high- and low-skill occupations, altering the division of labor between native-born and foreign-born workers, further differentiated along the lines of race, ethnicity, gender, and generation (e.g., Wright & Ellis, 1996, 1997; see also Kodras in this volume). Moreover, U.S. capitalists frequently have globalized without attention to the economic and social welfare of local communities and social groups (see Hollinger, 1995, p. 149; Natter & Jones, 1993).

While transnational corporations and migrants operate with seemingly scant regard for national borders, the same could scarcely be said for the constituency of voters who support Buchanan's anti-immigration stand. More generally, the voters who provided the critical support for the conservative 104th Congress rally in fierce defense of their country (and their view of its distinctive history, culture, territory, and equality under the law) while simultaneously being deeply suspicious of the state. This form of politics (Breuilly, 1994) occurs concomitantly with increasing transnational integration. Following Rouse (1995), I argue here that nationalism thus becomes dialectically linked to transnationalism.[2]

The modern state was integral—some would say "necessary" (Kearney, 1991)—to the development of capitalism. The Age of Empire (Hobsbawm, 1987) involved the bureaucratic, intellectual, and popular definition of colonizing states, as differentiated from other colonizers and the colonized. New state-based political and social formations stemming from the Reformation and the Enlightenment gradually replaced previous religious or dynastic orthodoxies (Anderson, 1983; Johnson, 1995), whereby communities had been built on ties of kinship, personal obligation, and fealty (Agnew, 1994). The modern state went hand in hand with the construction of absolute boundaries, inscribed both on territory and on people (Anderson, 1983; Sack, 1986). It follows that the "alien" (and the "illegal immigrant") is very much tied up with the rise of the modern state. It also follows that the evolution of state boundaries occurs in lock-step with the development of the modern state, and so too does nationalism. Breuilly (1994) makes the case emphatically that nationalism "is, above all and beyond all else, about politics and politics is about power. Power in the modern world is principally about control of the state" (p. 1).

If we are indeed entering a transnational era, control of the state is slipping. Transnationalism "challenges the rigid, territorial nationalism of the nation-state" (Lie, 1995, p. 304). Certain borders, notably the Mexican-U.S. border, are being replaced with "borderlands" of shifting and contested boundaries (Basch, Glick Schiller, & Szanton Blanc, 1994; Glick Schiller, Basch, & Szanton Blanc, 1992; Kearney, 1991; Martinez, 1994). Transnationalism, moreover, opposes the dogmatism and the binary logic of the nationalism—us vs. them, first world vs. third world, colonizer vs. colonized, and so forth. It also implies that we have entered a "postnational" age—one where the circulation of capital, goods, information, and people occurs with decreasing regard for national boundaries (Kearney, 1991, 1995; Szanton Blanc, Basch, & Glick Schiller, 1995).[3] Thus, we see the operation of the dialectic between transnationalism and state-based forms of nationalism.

In the mid-19th century, the United States justified the extension of its western boundary using nationalistic rhetoric encapsulated in the term *Manifest Destiny*. At that time, the rapid territorial acquisition occurred for many reasons, but undergirding the process lay innovations in transportation (e.g., railroad construction) and communication technologies (e.g., the telegraph). In the 1840s, as the economy grew apace, the U.S. border crossed over many peoples, including indigenous populations, peoples of mixed indigenous and Spanish heritage, and white settlers (e.g., Agnew, 1994). In the late 20th century, innovations in transportation and communication technologies are again linked with new economic expansion. Today, capital and labor cross international boundaries in search of higher returns, and the incorporation of new populations within the borders of the United States again has sparked nationalistic political action. We have no contemporary term to match Manifest Destiny as yet, but border crossing—transnationalism, if you like—be it in the form it took 150 years ago or the form it takes today, continues to be dialectically linked to nationalism.

■ To Die For

> I sometimes think that I might die while crossing, but I've just got to do it. (Mixtec [im]migrant quoted in Soguk, 1996, reflecting on traversing the U.S.-Mexican border)

> Tuesday, June 18, 1996, Tucson, Arizona. Five illegal aliens believed to have been crossing the desert in 108-degree heat in search of farm work were found dead from dehydration, heat exposure or both, U.S. Border

Patrol agents said. A man believed to be the group's lone survivor was found badly dehydrated and delirious Wednesday. The survivor told agents he had left the other men behind in a desolate area south of Arizona City. . . . Earlier this month, the bodies of four female illegal immigrants were discovered on ranch land near two south Texas immigration checkpoints. All apparently died of heat exhaustion. Copyright 1996 *The Associated Press.*

Territorial sovereignty is an indispensable attribute of independent nations; "the sacred soil" in whose defense true citizens will be prepared to give their lives. (Gottmann, 1973, p.15)

Many international migrations are the actions of desperate people. Individuals frequently endanger their lives making the crossing. This high stakes risk-taking has a long history. Death rates ranged as high as 20% crossing the Atlantic in the early 19th century (Handlin, 1973). Takaki reports similar figures for Asians crossing the Pacific two decades later (Takaki, 1989). Today, although most migrants enter without physical harm, many *pollos*—literally "chickens," the name given to illegal crossers—are beaten and robbed by their coyotes (people smugglers) or abandoned in remote deserts and locked railroad cars. Urrea (1993) reports that gangs from either side of the border attack illegals (and their coyotes) for sport. "[R]ape and gang rape are so common . . . as to be utterly unremarkable" (Urrea, 1993, p. 15).

Despite the risks of migration and the heavy human toll exacted during the crossing, immigrants to the United States have always moved back and forth between origin and destination communities (Bourne, 1916). The historical record brims with examples of such connectivity. These links have included not only the return of the disaffected (e.g., Shepperson, 1965) and the repatriation of the unwanted (Balderrama & Rodriguez, 1995), but also of those who returned home to buy land (e.g., di Leonardi, 1984) or businesses that ranged from small shops to steel works (Wyman, 1993) based on savings accumulated in the United States. Returnees also carried religious and political ideas, skills, and know-how (e.g., Cinel, 1991; Thomas & Znaniecki, 1918-1920; Wyman, 1993). Return rates for Europeans in the 1920s varied from 5 to 89% according to nationality (Wyman, 1993, p. 11). Bourne (1916) and Handlin (1956) both point out that some migrants moved between their place of birth and their homes in the United States several times. Bourne's (1916) "Trans-national America" also shows that some of the terms developed to understand our changed reality are not new either.

Ironically, before the broad-based U.S. immigration legislation of the 1920s, it was probably easier for immigrants to enter and exit the United States than it is today. For earlier immigrants to the United States, however, "re-migration" meant simply a return migration (Wyman, 1993). In other words, previous return international migrations were based on, first, an immigration, followed by a re-emigration. "This interruptive rhythm is a long way from the meter that characterizes the fluid functioning of linkages today. E-mail, phone, fax, jet aircraft, videos, and televisions operate to syncopate spatial interaction" (Mountz & Wright, 1996). These trends affect not only the wealthy but also those who lack many resources (see, for example, Kearney, 1991; Mountz & Wright, 1996; Rouse, 1991). The immigration, uprootedness, and absorption/assimilation of the past are now giving way to (im)migration (Rouse, 1995), hybrid cultures, and the indigestion of transnational migrants.

This backdrop of remarks about transnationalism, and transnational migration in particular, provides relief to, and perspective on, the landscape of national totemism. Nationalistic rhetoric and political action occur at a time when transnational processes pose a threat to political control of the state (Breuilly, 1994). The reassertion of control over the border area with Mexico through such policies as Operation Gatekeeper and Operation Hold-the-Line is part of a more general defensive action to control space that has been invaded by "foreigners" and "illegal aliens." Gated communities represent one aspect of this defensive reaction, but so too does the discourse on English as the official language in which the state seeks power to legislate identities and social practice (Kearney, 1991). The pervasiveness of public resistance to intrusion is perhaps best reflected in the breadth of political support for Patrick Buchanan in his effort to gain the Republican Party's presidential nomination and influence the Republican platform. Another indicator is the support for California's Proposition 187 ballot initiative, which gained support not only in metropolitan white enclaves, but also in a considerable number of Hispanic neighborhoods (Clark, 1996).

All the while, "illegal aliens" are being targeted, not their employers. These attacks on "illegal aliens" are so widespread and alarming that "legal aliens" in unprecedented numbers are now switching their immigration status from "permanent resident alien" ("green card" holders) to "citizen."[4] This logic also helps explain why the border and fence-jumping illegal aliens are targeted for action. It matters little that the majority of "illegal aliens" in the United States are here because they overstayed their visas rather than jumped a fence. It matters little that "illegal aliens" who work in the United States are employed by "legal aliens" or citizens (or in the case of the new INS building in Atlanta, by the INS itself, who employed several dozen "illegal aliens"). It matters a great deal to the nationalists that the INS seal the border (with Mexico) and thereby attempt to secure the nation and its identity. Violence at the border is very much bound up with sealing territorial boundaries. Connolly (1996) reminds us of the ambiguity associated with the derivation of territory. The *American Heritage Dictionary*, for example, states that the word *territory* derives from *terra,* meaning "land, earth, nourishment." The *Oxford English Dictionary* suggests an alternative derivation, *terrere,* meaning "to frighten, terrorize." A *territorium* is "a place from which people are warned." Connolly (1996) hypothesizes that these opposing derivations "continue to occupy territory today. To occupy a territory is to receive sustenance and to exercise violence" (p. 144).

In direct opposition to nationalism, Appaduri (1993) suggests it may be time to rethink patriotism; to redirect the material problems we face, such as the environment, race, drugs, and employment; to redefine "those social groups and ideas for which we would be willing to live—and to die" (p. 427). Michael Peter Smith (1994) makes largely the same point arguing for the "globalization of grassroots politics." The transnational is again set against the national. Rouse (1995) and Ong (1995) offer a slightly more nuanced reaction and ask us to think about the forces that underpin the national and the transnational. Rouse explores multiple sources of national power that include not only the state but also other actors. These ruling blocs not only rely on state political institutions but also exert influence through the mass media, churches, schools, and other institutions (cf. Gramsci, 1971). Ong (1995) contrasts (i) a multicultural nation of hybrids sustained by transnational links ranked in opposition against nationalism with (ii) the more subtle observation that Asian immigrants' concerns frequently prioritize free trade over human rights (cf. Appaduri, 1993). Using a different transnational perspective, she also maintains that the countries of origin of Asian immigrants challenge the U.S. ideal that fully fledged democracy and capitalism go together. Perspectives such as these lead to a contemplation of class interests (Rouse, 1995), the relationship between democratic freedoms and free trade (Ong, 1995), and, in the language of this book, the implications of transnationalism for a diverse society.

Transnational Migration, the State, and Diverse Societies

Transnationalism affects places and peoples in the United States unequally. NAFTA and other such agreements involve new strategies of investment and accumulation, benefiting certain societal groups more than others. Nationalism and protectionism appeal to those social groups who are, or believe themselves to be, disadvantaged or adversely affected by transnationalism and globalization. The state is intricately woven into this conflict between nationalist and transnationalist forces, simultaneously appealing to different classes, social groups, and individuals. The discourse of political campaigns and the funneling of funds to grow the INS are legitimating mechanisms that serve to appease some of the electorate, but at the same time the government allows and spurs accumulation on a global scale through policies such as GATT and NAFTA, thus lining the pockets of a select few social elites.

At one level of analysis, this characterization of transnationalism and class interests pits proletariat against bourgeoisie. But transnational processes have helped reshape class structure in the United States (Rouse, 1995). New sectors, reconfigured industries, and new types of work have emerged with a distinctively transnational signature. At the same time, older sectors have been "downsized" or are now absent from the United States altogether. These transitions extend beyond the economic.

> Hegemonic influence should be seen as involving not only efforts to influence coalition formation and to generate a broad consent to specific forms of coalitional rule, but also attempts to produce the kinds of subject deemed appropriate to prevailing economic and political arrangements. And the production of subjects should be viewed as addressing people's attitudes and aspirations regarding their relationship to work and their activities as consumers as much as the ways they conduct themselves as citizens. (Rouse, 1995, p. 396)

So while transnational forms of economic organization depend on geographically permissive technologies, they generate new expressions of social power and inequality in the United States.

Similarly, transnational migration confronts the nation, yet at the same time, reshapes it. Asian immigration to the United States—particularly from Southeast Asia (Ong, 1995)—presents a good example.

Asian immigration to the United States is very important in terms of its magnitude: Taken as a group, the numbers of Asian-origin, foreign-born persons declaring they "came to stay" in the United States exceeded the total number of "arrivals" from Mexico (U.S. Bureau of the Census, 1993). On the one hand, the influx of Asian peoples bolsters the image of the United States as the bastion of freedom and human rights and the conqueror of communism. (The same arguments and imagery also apply to immigrants and refugees from the former Soviet Union, countries of Eastern Europe, and Cuba.) On the other hand, Southeast Asians are also linked to a war in which the United States suffered defeat (as are Cuban immigrants), and their presence in the country serves as a symbol of U.S. military and moral weakness (Ong, 1995, p. 807).

Ong also points out that the economic success of trans-Pacific capitalism, originating in Singapore, China, Vietnam, South Korea, and the like, challenges not only the notion that the Pacific Rim is a region controlled by U.S. interests, but also the idea fundamental to U.S. nationalism that capitalism and democracy are intertwined. The knitting together of capitalism and democracy penetrates to the very center of U.S. nationhood, for if nothing else, U.S. national identity is linked to this ideology rather than a long-established territory or cultural uniformity. In the mid-19th century, many in civil society used Manifest Destiny to justify geographic and economic expansion. In the late 20th century, as the warp of capitalist institutions unravels from the woof of democratic tradition, "the transnational" unsettles "the national" in ways far more profound than sporadic confrontations along the Mexican-U.S. border or police chases between authorities and undocumented immigrants in suburban Los Angeles.

■ Conclusion

Nationalism associated with the modern state emerged in an era of mass communication, when human interaction superseded face-to-face contact. The imagined communities we refer to as nations are artifacts and as such will likely disappear at some point in time (Anderson, 1983). Transnational processes undermine national boundaries, complicate cultural compositions, and challenge ideological positions. Transnational processes also directly confront nationalistic rhetoric and totemism. But does it follow that transnationalism will spell the end for the nation?

Conceived as a nation, the United States is probably more important to the lives of immigrant workers today than it was 80 or 100 years ago (Hollinger, 1995). Not only is contemporary international migration more closely regulated and surveilled, but due to its recent worldwide expansion, today's immigrants are also more prepared for U.S. popular culture (Hollinger, 1995). Furthermore, once in the United States, newcomers can tap into political institutions and state-organized policies (such as affirmative action) that were not in place until 30 years ago.

Clearly, not everyone will rethink patriotism. Many individuals are still willing to die for the United States, and new citizens swear an oath to that effect. The question of whether transnationalism is bringing about the end of the nation is perhaps not the way to think about the transnational/national dialectic. As Anderson (1983) so forcibly established, all solidarities are social constructs; therefore, allegiance need not be to either the nation or the transnation. We should expect and support a diversity of allegiance, some of which could be national, some transnational. This could involve the obvious kinds of links between membership in international Greenpeace and supporting Ralph Nader's run for the U.S. presidency as a Green candidate, although the multiplicity and overlap of allegiances can be more complicated than that.

Let me conclude this essay with a particularly complex example that illustrates the role played by international migration in altering the transnational/national dialectic and state transformation—specifically, the relationship between transnational migration and attempts at Mexican political reform. Mexico sends more migrants to the United States than any other country. Mexicans living in the United States, however, have one of the lowest rates of naturalization of any immigrant group. That, of course, is changing, as the fear of reduced opportunities for noncitizens and citizenship campaigns by Latino organizations have combined to produce record rates of Mexican naturalization in 1995. Whether or not Mexican-Americans become a significant political force in the near future, however, depends as much on events outside the United States as inside. In April 1996, Mexico's Institutional Revolutionary Party (PRI) and the Democratic Revolutionary Party (PRD) reached agreement on reforms that would effect more representative and fairer elections in Mexico. These reforms include the proviso that Mexican citizens living abroad be allowed to vote in Mexico's presidential elections. This means that the 5 million or so adult Mexican immigrants residing in the United States would be eligible to vote in Mexican elections—as long as they remained Mexican citizens. Not only does this raise the possibility of Mexican presidential campaigns becoming transnational (because over 10% of the Mexican electorate live in the United States), but the proposed law could profoundly affect rates of naturalization among Mexicans in the United States. Also, because Mexican law enables ex-citizens to regain citizenship with relative ease, the proportion of Mexican-American citizens in the United States might actually drop. Alternatively, Mexican-Americans with U.S. citizenship might also become dual citizens and become part of a transnational electorate, negotiating political affinities between the PRI, the PAN, and Republicans, or the PRD and the Democrats! Thus the national and the transnational intersect in interesting ways, requiring us to think about our own solidarities, the ramifications of boundary drawing and redrawing, and the implications of all this for a diverse society.

NOTES

1. Thanks go to Jan Kodras and Colin Flint for their comments and their detailed reading of earlier versions of this paper. Conversations with Alison Mountz, Frances Ufkes, Jeff Garneau, and Sheila Culbert at Dartmouth, as well as with Mark Ellis and Joe Nevins at UCLA, helped inform these remarks. The section on (im)migration draws heavily on Mountz and Wright (1996).

2. This discussion of nationalism and transnationalism requires explicit definitions. I refer here to a particular form of nationalism specifically known as state-nationalism, which is a conscious cohesiveness of interest and purpose defined by the history of a country and demarcated along territorial boundaries of the state. This contrasts with ethnic-nationalism, whereby cohesiveness is defined according to the historical traditions of a particular cultural group without regard for state boundaries. In the case of Japan, ethnic and state forms of nationalism are the same, because the spatial extent of the cultural group and the boundaries of the state for the most part coincide. But in the United States and the many other countries containing diverse ethnic populations, the two forms of nationalism are distinct. I refer in this discussion to state-nationalism. I also use the expression transnationalism advisedly (Kearney, 1995). The terms *globalization* and *transnationalization* infect much current thinking about society, yet they are not the same. As Kearney (1995) argues (and others, for example, Basch et al., 1994; Glick Schiller et al., 1992) global processes are largely decentered from specific countries, while transnational processes are rooted in and transcend country boundaries. The difference is important, because the transnational calls attention to the state by directly opposing the cultural-political-economic projects of "nationalists." The transnational directly conflicts with the jurisdictional power of states. Globalization is more abstract, less institutional, less intentional—associated with, for example, technological change and the new global information economy. It is more universal and more impersonal, removed from the daily lives of people (Kearney, 1995).

3. The "post-national" yields yet another "post prefix" that aligns with Barnes's (1996) four main characterizations of post-prefixed anti-Enlightenment thinking.

4. The INS received about one million applications for citizenship in 1995, quintuple the level 5 years ago. Another one million applications are expected in 1996.

REFERENCES

Agnew, J. (1994). The territorial trap: The geographical assumptions of international relations theory. *Review of International Political Economy, 1*, 53-80.

Anderson, B. (1983). *Imagined communities*. New York: Verso.

Appaduri, A. (1993). Patriotism and its futures. *Public Culture, 5*, 411-429.

Balderrama, F. E., & Rodriguez, R. (1995). *Decade of betrayal: Mexican repatriation in the 1930s*. Albuquerque: University of New Mexico Press.

Barnes, T. (1996). *Logics of dislocation*. New York: Guilford.

Basch, L. G., Glick Schiller, N., & Szanton Blanc, C. (1994). *Nations unbound: Transnational projects, postcolonial predicaments, and deterritorialized nation-states*. Langhorne, PA: Gordon and Breach.

Bourne, R. (1916). Trans-national America. *Atlantic Monthly, 118*, 86-97.

Bratsberg, B. (1995). Legal versus illegal U.S. immigration and source country characteristics. *Southern Economic Journal, 61*, 715-727.

Breuilly, J. (1994). *Nationalism and the State* (2nd ed.). Chicago: University of Chicago Press.

Buchanan, P. (1996). Stump speech. New Hampshire Primary.

Cinel, D. (1991). *The national integration of Italian return migration, 1870-1929*. New York: Cambridge University Press.

Clark, W. A. V. (1996, April). *Migration and reaction: Interpreting the vote for Proposition 187*. Paper presented at the Annual Meeting of the Association of American Geographers, Charlotte, NC.

Connolly, W. E. (1996). Tocqueville, territory, and violence. In M. J. Shapiro & H. R. Alker (Eds.), *Challenging boundaries* (pp. 141-164). Minneapolis: University of Minnesota Press.

di Leonardi, M. (1984). *The varieties of ethnic experience: Kinship, class, and gender among California Italian Americans*. Ithaca, NY: Cornell University Press.

Glick Schiller, N., Basch, L., & Szanton Blanc, C. (Eds.). (1992). Towards a transnational perspective on migration: Race, class, ethnicity, and nationalism reconsidered. In *Annals of the New York Academy of Sciences* (Vol. 645). New York: New York Academy of Sciences.

Gottmann, J. (1973). *The significance of territory*. Charlottesville: University Press of Virginia.

Gramsci, A. (1971). *Selections from the prison notebooks of Antonio Gramsci*. (Q. Hoare & G. N. Smith, Ed. and Trans.). London: Lawrence & Wishart.

Handlin, O. (1956). Immigrants who go back. *Atlantic Monthly, 158*, 70-74.

Handlin, O. (1973). *The uprooted* (2nd ed.). Boston: Little, Brown.

Higham, J. (1988). *Strangers in the land: Patterns of American nativism 1860-1925*. New Brunswick, NJ: Rutgers University Press.

Hobsbawm, E. J. (1987). *The age of empire, 1875-1914*. New York: Pantheon.

Hollinger, D. A. (1995). *Postethnic America—Beyond multiculturalism.* New York: Basic Books.
Johnson, N. C. (1995). The renaissance of nationalism. In R. J. Johnston, P. J. Taylor, & M. Watts (Eds.), *Geographies of global change: Remapping the world at the end of the twentieth century* (pp. 97-110). Oxford, UK: Basil Blackwell.
Kearney, M. (1991). Borders and boundaries of state and self at the end of empire. *Journal of Historical Sociology, 4,* 52-74.
Kearney, M. (1995). The local and the global: The anthropology of globalization and transnationalism. *Annual Review of Anthropology, 24,* 547-65.
Lie, J. (1995). From international migration to transnational diaspora. *Contemporary Sociology 24,* 303-306.
Martinez, O. J. (Ed.). (1994). *Border people: Life & society in the U.S.-Mexico borderlands.* Tucson: University of Arizona Press.
Mountz, A., & Wright, R. (1996). Daily life in the transnational migrant community of San Agustin, Oaxaca, and Poughkeepsie, New York. *Diaspor, 6.*
Natter, W., & Jones, J. P. (1993). Pets or meat: Class, ideology and space in "Roger and Me." *Antipode, 25,* 140-158.
Ong, A. (1995). Southeast Asian refugees and investors in our midst. *Positions, 3,* 806-13.
Rouse, R. (1991). Mexican migration and the social space of postmodernism. *Diaspora, 1,* 8-23.
Rouse, R. (1995). Thinking through transnationalism: Notes on the cultural politics of class relations in the contemporary United States. *Public Culture, 7,* 353-402.
Sack, R. (1986). *Human territoriality: Its theory and history.* Cambridge, UK: Cambridge University Press.
Shepperson, W. S. (1965). *Emigration & disenchantment: Portraits of Englishmen repatriated from the United States.* Norman: University of Oklahoma Press.
Smith, M. P. (1994). Can you imagine? Transnational migration and the globalization of grassroots politics. *Social Text, 39,* 15-34.
Soguk, N. (1996). Transnational/transborder bodies: Resistance, accommodation, and exile in refugee and migration movements on the U.S.- Mexican border. In M. J. Shapiro & H. R. Alker (Eds.), *Challenging boundaries* (pp. 285-326). Minneapolis: University of Minnesota Press.
Szanton Blanc, C., Basch, L., & Glick Schiller, N. (1995). Transnationalism, nation-states, and culture. *Current Anthropology, 36,* 683-6.
Takaki, R. (1989). *Strangers from a different shore: A history of Asian Americans.* New York: Little, Brown.
Thomas, W. I., & Znaniecki, F. (1918-1920). *The Polish peasant in Europe and America: Monograph of an immigrant group,* Vols. 1-5. Chicago: University of Chicago Press.
U.S. Bureau of the Census. (1993). *1990 census of population public-use micro-data sample.* Washington, DC: Government Printing Office.
Urrea, L. A. (1993). *Across the wire: Life and hard times on the Mexican border.* New York: Anchor.
Wright, R., & Ellis, M. (1996). Immigrants and the changing racial/ethnic division of labor in New York City, 1970-1990. *Urban Geography, 17,* 317-353.
Wright, R., & Ellis, M. (1997). Nativity, ethnicity, and the evolution of the intra-urban division of labor in Los Angeles. *Urban Geography, 18.*
Wyman, M. (1993). *Round-trip to America: The immigrants return to Europe, 1880-1930.* Ithaca, NY: Cornell University Press.

12

Education Policy and the 104th Congress

FRED M. SHELLEY

The 104th Congress was the scene of vigorous and often acrimonious debate over numerous domestic policy issues. The Republican majority introduced proposals that would result in dramatic downsizing and restructuring of many services, including welfare, crime prevention, and environmental protection. Although public education is a highly controversial service, the Republican majority in the 104th Congress did not attempt a systematic overhaul of education policy.

Why has the Republican majority made no concerted effort to enact broad, fundamental changes in the relationships between government and education? The purpose of this chapter is to explore this question. The chapter is divided into three sections. First, the debate within the Republican Party over the proper role of government is discussed. The second section presents the historical context of debate over educational policy. The final section addresses present and possible future relationships between education policy, state devolution, and contemporary politics in the United States.

■ The Republican Dilemma

After the 104th Congress was seated in January 1995, the Republican majority attempted to capitalize on what its leaders regarded as a

---

AUTHOR'S NOTE: The helpful comments of Arlene Shelley, Julie Tuason, Karin Ascot, Jan Kodras, Lynn Staeheli, Colin Flint, and John Paul Jones on earlier drafts of this chapter are gratefully acknowledged.

mandate from the American public to restructure domestic policy. Although the relatively uncontroversial proposals contained within the *Contract with America* were supported with little opposition by most Republican members of Congress, unity within the Republican ranks began to unravel once Congress began to address more controversial issues. For example, more than 50 Republican members of the House of Representatives voted with most of the Democrats to defeat a bill to weaken the regulatory authority of the Environmental Protection Agency.

Why did the Republican leadership fail to deliver majorities in favor of its more controversial proposals? This failure may be due to the fact that the Republican Party consists of individuals who hold moral and philosophical views that are difficult to reconcile, if not fully incompatible. This incompatibility within the Republican Party results from the general tendency for Democrats to regard government as a vehicle to solve problems, while Republicans regard government as *the* problem.

During the New Deal, the Democratic Party, under the leadership of Franklin Delano Roosevelt, became associated with support for expanded federal authority (Leuchtenberg, 1993). Ever since, Republicans have debated how to react to Democratic proposals associated with increasing the size, scope, and authority of the federal government. Commenting on the divisive Republican presidential primaries of 1964, for example, journalist Theodore H. White (1965, p. 62) stated that "[t]he Republicans' impossible dilemma is that they have never sorted out properly what it is that government should do and should not do—and at what level."

The Republican ranks include both "moderates" and "conservatives." Moderates generally call for accepting the basic philosophy of increased federal responsibility, although this acceptance is tempered by calls for more efficient, cost-effective, and "businesslike" management of governmental programs. Conservatives, on the other hand, have argued for eliminating federal programs and reducing the scope of federal authority. For the past half century, dissension between moderates and conservatives has characterized battles for Republican Party presidential nominations. Moderates prevailed in 1952 when Dwight D. Eisenhower defeated Robert Taft and in 1976 when Gerald Ford defeated Ronald Reagan. Conservatives prevailed, however, with the nomination of Barry Goldwater over Nelson Rockefeller in 1964 and with Ronald Reagan's victory over George Bush in 1980.

The debate between moderates and conservatives over the party's presidential nomination in 1996, therefore, was by no means unprecedented. The eventual nominee, Robert Dole, represented the moderate wing of the party, whereas several of his major primary opponents represented the conservative viewpoint. Yet Dole's conservative opponents disagreed vigorously among themselves. Pat Buchanan, for example, advocated opposition to abortion, restrictions on immigration, rejection of the North American Free Trade Agreement (NAFTA) and the General Agreement on Tariffs and Trade (GATT), and increased protective tariffs. In effect, Buchanan's supporters argued for reducing the integration of the United States into a multicultural global economy. As Buchanan said in several campaign speeches, "Let's cancel the New World Order."

Steve Forbes, on the other hand, supported NAFTA and GATT and called on Republicans to offer vigorous support for free trade. He argued that a "flat," as opposed to progressive, tax system would stimulate economic activity. Forbes advocated increased economic activity and international trade and argued for expanded commitment to improving America's position in the global economy. In contrast to Buchanan, Forbes did not take strong positions on social issues.

The differences between Buchanan's and Forbes' views on economic and social policy illustrates the depth of the Republican dilemma in a post-Cold War world. Because capital flows into the most profitable regions or sectors of an economy, capitalism rewards innovation and diversity. Yet Buchanan's philosophy of increased social control based on Judeo-Christian ethical standards would, if implemented, be likely to stifle the cultural and economic diversity necessary to sustain growth in a capitalist economy. Capitalism depends on cultural diversity. On the other hand, government regulation of moral issues is likely to inhibit cultural diversity. The inhibition of diversity is likely to stifle the possibility of sustained economic growth associated with successful competition within the world economy.

The Republican Party today includes strong supporters of improved American competitiveness within the world economy as well as advocates of increased government regulation of social and moral issues. The depth of disagreement within the Republican ranks concerning the relative importance of economic policy and moral issues illustrates the degree to which conservative positions on these issues are incompatible with one another. Because education is so intimately associated with questions of fundamental values, and because education is the single policy arena most closely associated with local control, the Republican majority in Congress has yet to initiate efforts to promote anything more than piecemeal reform of public schooling.

■ Education and the Republican Dilemma

Throughout U.S. history, the provision of public education has been an issue of considerable controversy and political debate. Public schooling as we know it today came into being during the 19th century, when most states enacted laws providing for the establishment of tax-supported, public common schools (Cubberley, 1934; Kaestle, 1983). Laws were passed establishing local districts and empowering them to levy taxes in order to maintain public schools. Local education authorities "were granted almost complete freedom in carrying out state laws mandating the creation of schools" (Reynolds & Shelley, 1990).

The common school movement flourished between the 1820s and the 1850s, coinciding with the beginning of the Industrial Revolution in the United States and with the arrival of the first wave of non-Anglo-Saxon immigrants (Reynolds & Shelley, 1990). Supporters of the common school movement believed that class conflict associated with these changes "could be avoided by mixing the children of the rich and the poor in the same schoolhouse and using the common school to provide equality of opportunity" (Reynolds & Shelley, 1990, p. 112). Moral and political education were emphasized in the common schools, which became an important vehicle in the creation of a political community. As Paris (1995) has indicated, common schools would not only provide "the basic literacy essential for the political, economic, and social life of all citizens," but they would "help forge a social bond by providing common moral and political understandings to otherwise different individuals and groups" (p. 62). In practice, however, the ideology taught in most common schools was "a republicanism that emphasized the need for public obedience rather than public participation" (Reynolds & Shelley, 1990, p. 113).

During the first three decades of the 20th century, however, local control over common schools came to be replaced by control over public education by authorities at higher levels of government (Tyack, 1974). The reform of public education was a high priority of the Progressive movement. The Progressive reformers "shared a conviction that education was the prime means of directing the course of social evolution" (Tyack & Cuban, 1995, p. 17). They proposed to establish uniform educational standards and grant more authority to educational professionals.

In response to the Progressives' proposals, legislatures "increasingly standardized schools across the nation according to the model of a modern school proposed by the policy elite" (Tyack & Cuban, 1995, p. 19). States began to require teachers to hold college degrees, to establish minimum school day and school year lengths, and to impose uniform high school graduation requirements. The Progressives also argued for depoliticizing the public schools. For example, they advocated the replacement of elected school boards and superintendents by civil servants selected on the basis of merit and professional qualifications. As these reforms were enacted into law by various states, local control of education passed into history (Reynolds & Shelley, 1990). By the time of World War II, American public education was administered primarily by reformers who "shared a common faith in 'educational science' and in lifting education 'above politics' so that experts could make the crucial decisions" (Tyack & Cuban, 1995, p. 17).

In 1957, the Soviet Union launched its Sputnik satellite. Americans began to express concern that their children were receiving inadequate instruction in science and mathematics. Influential critics claimed that American public schools were antiquated and outmoded (Conant, 1959). Education came to be regarded as a national problem. The Sputnik crisis coupled with federal intervention in school integration following the Supreme Court's decision in *Brown v. Board of Education* in 1954 resulted in more direct federal involvement in education. Education-related controversies have received substantial attention on America's political agenda ever since.

Prior to the 1950s, most Americans believed that public schools in the United States were of high quality. A Gallup poll taken in 1946, for example, indicated that 87% of American parents were satisfied with the schools that their children attended (Tyack & Cuban, 1995, p. 13). By the 1970s, however, Americans' confidence in their public schools had begun to erode. At the same time, many Americans began to resent increased federal involvement in education. This resentment was fueled by unpopular mandatory school busing programs, declining standardized test scores, reports of violence and drug abuse in schools, high dropout rates, and continued increases in the cost of public education relative to the cost of living.

The Republican Dilemma and Its Effects on Education Policy

As Americans became more disenchanted with federal education policy, the positions of the two major parties concerning educational policy began to diverge. Prior to the 1980s, in contrast, "the educational policies of Republicans and Democrats . . . tended to move together in tandem over time, often following public opinion as much as leading it" (Tyack & Cuban, 1995, p. 45). More recently, however, the positions of educational professionals and major teachers' organizations have come to be identified with the Democrats, leaving the Republicans in opposition. Proposals to institute school choice and return voluntary prayer to public schools—both forms of dismantling government bureaucracy and regulation—have come to be associated with the Republicans.

Over the past two decades, numerous scholars, educators, journalists, business executives, and politicians have issued calls for reforming America's public schools. With so much discussion of how education should be reformed, and with increasing divergence between the two major parties over education as a political issue, why has the Republican majority not undertaken a concerted effort to institute large-scale, fundamental reforms in public education? It is argued here that such full-scale reform has not been addressed explicitly in the political arena *because* of the Republican Party's dilemma. Reforms promulgated by supporters of increased competitiveness in the global economy are unsatisfactory to those supporting tighter social controls, and vice versa. In this section, it is argued that any effort to propose a significant overhaul of the educational system will force reform proponents and opponents to examine and debate fundamental values—and meaningful debate over values, as we have seen, is likely to rupture the fragile Republican coalition.

The fragility of the Republican coalition as applied to education-related issues is illustrated with reference to several proposed educational reforms that have been associated with or are consistent with the philosophical views of various factions within the Republican coalition. This point is illustrated with reference to four frequently debated reform proposals: improving the competitiveness of students in U.S. schools; financing education; school choice; and increasing local control.

Numerous observers have claimed that U.S. public education is "inferior" to public education in Japan, Germany, and other countries.

Those articulating such claims point to evidence that the performance of U.S. students on standardized tests is consistently weaker than that of pupils elsewhere. On this view, failure to improve the performance of U.S. students will contribute eventually to the declining competitive position of the United States in the global economy.

In analyzing why foreign educational systems are "better," proponents of change point to several supposed shortcomings of contemporary education in the United States. Some point to evidence that children in other countries spend more time in the classroom. For example, Japanese schoolchildren attend school for an average of 220 days a year as compared to 180 days in the United States (Schlossstein, 1989). Moreover, the school day in Japan and other countries is longer than in the United States, and a higher percentage of the students' time in the classroom is devoted to academic subjects.

Calls for increasing the competitiveness of American schools and students have been accompanied by proposals to establish and maintain national educational standards. In many countries, high school students are required to pass national examinations or meet national curricular achievement standards before graduation.

Should American education be restructured along these lines? The debate over educational reform draws from the same sentiments as the debate over state restructuring generally; much of it revolves around the costs and benefits of devolution, dismantling, and privatization.

Although many have argued for longer school days and school years, increased emphasis on academic subjects, and more rigorous performance standards, most Americans are skeptical of reforms that would by definition be associated with increased centralization and that would run counter to the ideology of devolution and local control. The Japanese educational system is accompanied by a highly centralized bureaucracy very different from the decentralized structure of education in the United States. Japanese education "is tightly controlled and regulated by a powerful Ministry of Education; course offerings, textbooks, teachers' salaries and even a school's physical plant are under its supervision" (Kennedy, 1993, p. 139).

Although some educational professionals have called for similar efforts in the United States, any proposal to restructure education in the United States along the lines of education in Japan or other more centralized systems would meet with fierce resistance among most Americans. Local control remains a powerful ideology shaping American educational policy (Reynolds & Shelley, 1990). As a result, the education function is already a devolved one in the most fundamental and historic of ways. Not only would proposals to impose centralized, national educational standards be seen as violating the ideology of local control, but the proposals would be opposed by those who are skeptical about the desirability of increased integration of the United States into the global economy. Those Americans who argue that public schools indoctrinate children with values incompatible with their own would naturally oppose efforts to increase the length of time children are exposed to these undesirable values in the public schools.

This point is reinforced by the fact that recent efforts to develop national educational standards have met with failure. In 1989, for example, a blue-ribbon commission appointed by President George Bush empowered committees from various academic disciplines to develop national standards in their respective fields. The standards developed in history were regarded by many as inconsistent with "traditional" American values; and in 1995, large, bipartisan majorities in both houses of Congress voted to reject them.

Another issue of considerable controversy in contemporary U.S. public education is that of school choice. To what extent should parents be granted the right to choose what school their children attend? During the 19th century, of course, all children within the boundaries of any given school district, regardless of social or economic status, attended the same school. This philosophy, which was central to the common-school movement, was unaffected by the reforms of the 20th century, and parents living within any given school district have seldom had the opportunity to choose which school their children attended.

In recent years, the idea that school officials within districts have the authority to assign children to particular schools has come under fire for several reasons. Following the Supreme Court's declaration in *Brown v. Board of Education* (1954) that school segregation was unconstitutional, some courts and school districts began to institute mandatory school busing programs. Busing became an inflammatory political issue in the 1960s and 1970s, especially in those areas where children of affluent parents were assigned to attend schools in poor and minority-dominated neighborhoods.

Suburbanization contributed to the controversy over school choice. Some parents responded to actual or proposed busing programs by moving from central cities to suburbs located in "better" school districts. This "white flight" was encouraged by the Supreme Court's decision in *Milliken v. Bradley* (1974), which restricted the right of any

Education Policy and the 104th Congress

state to impose mandatory cross-district busing to achieve metropolitan-wide school desegregation. Encouraged by the busing controversy and increasing evidence of poor quality in urban schools, white flight exacerbated an already growing disparity in school funding between impoverished central cities and their affluent suburbs (Shelley, 1994).

Meanwhile, the traditional separation between public and private schooling came to be questioned. The common-school movement, of course, was founded on the philosophy of mixing students of all backgrounds to be taught a common curriculum. Advocates of common schooling expressed hostility to private schooling, and this hostility was exacerbated when Roman Catholic immigrants established church-sponsored parochial schools "to maintain their traditions and to insulate themselves from the Protestant bias prevalent in common schools" (Kahne, 1996, p. 100). In order to discourage private schooling, advocates of the common school movement pushed for laws prohibiting the use of public funds for private and parochial schools and requiring them to adhere to certification standards imposed by the state.

Over the past two decades, in contrast, the sharp distinction between public and private schools has become blurred. Presidents Reagan, Bush, and Clinton endorsed the idea of allowing public funds to be used for private and parochial schools (Kahne, 1996); and Dole announced support for this proposal during his campaign for the Republican nomination for president in 1996. It has been proposed that parents be granted tuition vouchers by the state, with these vouchers to be used to "purchase" education at public, private, or parochial institutions of the parents' choice. These proposals for greater privatization of the educational system have been accompanied by calls to allow greater freedom for parents to choose public schools within or between districts (Chubb & Moe, 1990). Unlimited school choice was enacted into law in Minnesota in the late 1980s; by 1993, proposals to allow school choice had been passed or introduced in 34 of the 50 states (Tucker & Lauber, 1994).

The debate over vouchers and school choice has revolved around several considerations and thus reflects the themes in the debate over state restructuring generally. Supporters claim that a voucher system coupled with parental choice will improve the quality of schools because schools will be forced to compete to attract critical masses of students. Chubb and Moe (1990) claimed that school choice would promote higher levels of academic achievement, better quality instruction, and more efficient and cost-effective education. In general, advocates on both sides of the school debate emphasize "equity, efficiency, and excellence" (Kahne, 1996, p. 99).

Less attention has been paid, however, to the relationships between vouchers and school choice proposals and their effects on the formation of political communities. If school choice and voucher programs are instituted, the traditional role of schools in promoting common moral and political values will be reduced, if not eliminated. Such a prospect has been regarded as troubling by influential supporters of the Republican agenda. William Bennett, who served as Secretary of Education under Reagan, stated that "there are values that all American citizens share and that we should want all American students to know. . . . The explicit teaching of these values is the legacy of the common school, and it is a legacy to which we must return" (Bennett, 1992, p. 58).

A further illustration of the difficulties associated with any potential Republican effort to reform public education is provided by examining the debate between proponents of increased multiculturalism and proponents of renewed emphasis on Western cultural norms. Throughout U.S. history, the curricula taught in common schools have reflected the cultural values dominant at any given point in time. During the 19th century, instruction in most common schools reflected a moralistic, Protestant, and Whiggish bias. Indeed, an examination of textbooks used by pupils in the 19th-century United States showed that typical books contained laudatory references to Federalists such as George Washington and Alexander Hamilton, in contrast to less detailed and more critical treatment of prominent Democrats such as Thomas Jefferson and Andrew Jackson. Indeed, the American Revolution itself was often described in such textbooks as a "conservative" revolution led by an elite consisting of men of wealth and property oriented to upholding law and order (Elson, 1964).

Curricular bias continued unabated during the 20th century. During the 1960s, many observers began to argue that curricula emphasized the achievements of white males at the expense of women, ethnic minorities, and people of color. Since that time, of course, curricular restructuring has emphasized multiculturalism, with more emphasis placed on women and minority group members in past and present U.S. society.

Today, many Americans believe that these efforts have gone too far and that they overemphasize multiculturalism. Many supporters of the Republican majority are identified with opposition to multiculturalism. Yet many of the same persons who oppose multiculturalism call for limited government and renewed emphasis on local, as opposed to federal, management of services. On this view, public schooling is inadequate because the local control that had once characterized educational governance has been usurped by governments at the state and federal levels.

Can local control be restored while multiculturalism is suppressed? In some inner-city communities, for example, local officials have developed programs that emphasize African American culture and history. Critics of such programs complain that insufficient attention is paid to traditional Western values in such curricula. Yet a fine line separates local control from local empowerment. Can proponents of local control criticize officials in minority-dominated communities who have exercised local control to develop multicultural curricula, even if these curricula are not based on the Anglo-American cultural tradition? Under a system of school choice, what is to prevent parents who wish their children to be educated in non-traditional settings from establishing such schools? More generally, can school reforms based either on increasing global competitiveness or on emphasizing Judeo-Christian cultural values be implemented without expanding, rather than reducing, the role of the federal government in public education? Resolving these contradictions between privatization, dismantling, and devolution is difficult, if not impossible. Yet unless they are resolved, it is unlikely that the Republican majority in Congress will address educational policy comprehensively.

■ Conclusion

Current debates over education and educational policy illustrate the fundamental dilemma faced by those who advocate a dramatic restructuring of the role of government in American life. Many who believe that the federal government currently fails to represent what they consider to be "American" values have supported the Republican majority in the 104th Congress. Because of the fundamental contradictions underlying the relationship between capitalism and social control, however, leaders and supporters of the Republican majority have not made a concerted effort to establish a meaningful agenda to restructure public education in the United States. Unless the Republican Party is able to resolve this fundamental dilemma, it is unlikely that full-scale, as opposed to piecemeal, educational reform will be given a high priority on the political agenda of the United States in the foreseeable future.

REFERENCES

Bennett, W. J. (1992). *The de-valuing of America: The fight for our culture and our children.* New York: Summit Books.
Brown v. Board of Education of Topeka, Kansas, 347 U.S.483 (1954).
Chubb, J., & Moe, T. (1990). *Politics, markets and America's schools.* Washington, DC: Brookings Institute.
Conant, J. (1959). *The American high school today: A first report to interested citizens.* New York: McGraw-Hill.
Cubberley, E. (1934). *Public education in the United States* (2nd ed.). Boston: Houghton Mifflin.
Elson, R. M. (1964). *Guardians of tradition.* Lincoln: University of Nebraska Press.
Kaestle, C. (1983). *Pillars of the republic: Common schools and American society, 1780-1860.* New York: Hill and Wang.
Kahne, J. (1996). *Reframing educational policy.* New York: Columbia University, Teachers College Press.
Kennedy, P. (1993). *Preparing for the twenty-first century.* New York: Random House.
Leuchtenberg, W. F. (1993). *In the shadow of FDR: From Harry Truman to Bill Clinton.* Ithaca, NY: Cornell University Press.
Milliken v. Bradley, 418 U.S. 717 (1974).
Paris, D. B. (1995). *Ideology and educational reform: Themes and theories in public education.* Boulder, CO: Westview.
Reynolds, D. R., & Shelley, F. M. (1990). Local control in American public education: Myth and reality. In J. E. Kodras & J. P. Jones III, *Geographic dimensions of U.S. social policy* (pp. 107-133). London: Edward Arnold.
Schlossstein, S. (1989). *The end of the American century.* Chicago: Congdon and Weed.
Shelley, F. M. (1994). Local control and financing of education: A perspective from the American state judiciary. *Political Geography, 13,* 361-376.
Tucker, A. M., & Lauber, W. F. (1994). *School choice programs: What's happening in the states.* Washington, DC: Heritage Foundation.
Tyack, D. (1974). *The one best system.* Cambridge, MA: Harvard University Press.
Tyack, D., & Cuban, L. (1995). *Tinkering toward utopia: A century of public school reform.* Cambridge, MA: Harvard University Press.
White, T. H. (1965). *The making of the president 1964.* New York: Atheneum.

13

Environmental Policy and Government Restructuring

MARVIN WATERSTONE

The Republican *Contract with America,* consistent with other rhetorical flourishes by the actors in both of the major U.S. political parties, seeks to reconfigure first, the dominant discourse regarding the appropriate functions of government, and second, how those functions still deemed legitimate ought to be carried out. Although the Contract with America has lost some of its visceral impact (and putative public appeal) in the intervening years since its deployment, it is still a useful emblem of a particularly powerful component of contemporary U.S. politics. Underlying the specific language of the Contract itself are contestations over issues that affect the everyday life of many citizens. In fact, in interesting ways, the Contract (and the broader political currents of which it is one discursive manifestation) attempts to redefine such fundamental notions as citizenship and the role of the state in society. In its attempts to control the discourse over such normative matters as "appropriate" regulation, "acceptable" risks, and "public" versus "private" property rights, the Contract goes a long way in endeavoring to define who participates in such matters and in what ways. In some very fundamental ways, this is a definition of what constitutes citizenship in a democracy. Through these discursive strategies, the Contract is also attempting to reconfigure the relationship between the state and civil society.

Nowhere are these attempts clearer in the Contract (and in the continuing legislative and discursive activities that have ensued since the 1994 elections) than in the myriad ways in which it essays to alter fundamental policies about the environment and natural resources. When dealing with environmental and resource issues, the architects of the Contract and its underlying politics have had to be especially cognizant of the need to preserve the legitimacy of the state while simultaneously furthering their view of the state's role in aiding the accumulation of capital. Often these two goals can be seen as antithetical, and balancing these pursuits has influenced the strategies used for restructuring the U.S. environmental regulatory apparatus. Of course, assisting this effort has been the ability of Republican representatives (led by Newt Gingrich) to control matters of public discourse over these issues to an unprecedented degree. The architects of the Contract have shown themselves to be quite adept at what one author has termed "the politics of perception" (Bruck, 1995).

In this chapter I will survey these proposed restructurings, but will also put them into context by examining the apparatus that is being dismantled, exploring the reasons for the present configuration of environmental policy (paying particular attention to the issues of federalism), and finally, by addressing the broader issues at stake in this debate.

■ What Is Being Restructured?

In order to understand what the current recommendations entail, it is necessary to examine the present structure of environmental protection and resource management. It is also important to ascertain why this structure came into being in the first place so that the implications of breaking down these arrangements can be illuminated.

The Present Structure of U.S. Environmental Policy

Most environmental problems represent a management conflict, which results in a challenge to finding an optimal solution, especially within a federalist governmental structure (Strohbehn, 1982). Because many environmental problems are externalities (i.e., market failures), it is frequently (though certainly not universally) acknowledged that they require governmental regulatory action rather than market approaches to resolve them. At the same time, many environmental problems involve concerns that differ significantly from one locality to another, making national (i.e., one-size-fits-all) solutions appear inappropriate or inefficient, particularly from an economic standpoint. This has resulted in a great deal of controversy over what scale of government is best equipped to deal with environmental issues (or indeed whether governmental venues are appropriate at all).

It should be clear that when both the costs and benefits of an issue are national in scope, there is little question that federal decision making is appropriate. It is equally clear that when the costs and benefits are both local or statewide in scope, again there is little question about jurisdiction. However, when the costs are all borne, or are largely borne, by a small group of people in one area, while the benefits are nationwide, or vice versa, problems of jurisdiction do arise (Smith, 1982). This is frequently the situation with environmental issues and is the basis for the federalist approach to such matters that has evolved over time in the United States.

A rudimentary sketch of the present system would look like this:

Problem→program design→compliance monitoring and enforcement→effect on problem→feedback

Of course, this description is highly schematic and simplified, and there remain many interesting questions regarding problem definition and the almost total lack of feedback, but here I am mainly concerned with what level of government has control over what parts of the policy process. As matters stand currently, this approach consists mainly of goals and objectives for environmental quality (i.e., problem definition) being determined at the national level, with control over matters of monitoring and enforcement turned over to state and local governments (with economic incentives and hierarchical supervisory safeguards by the federal government built into the system). This pretty much accords with the preferences of the major players. For example, industrial actors frequently favor federal control over goals, objectives, and standards—because this provides uniformity—but local or state control over implementation and enforcement—where these actors believe they have more influence. The environmental community (at least as represented by the major environmental organizations) generally favors national regulations for consistency and to prevent arena-shopping, but sometimes wants local control over issues such as the siting of noxious facilities.

Rationale for the Present System

Over time, a number of justifications have been offered for federal involvement in environmental policy; and as these have gained acceptance and become part of the legal and regulatory framework, a role for state and local governments has also evolved in the environmental arena. By examining this framework and the arguments underlying it, it will be possible to identify some of the implications involved in the current round of proposed restructuring. In no particular order, the chief arguments for a federal role in environmental protection have included: (a) preventing one state from polluting another; (b) protecting uniquely national areas set aside because of environmental significance for the nation as a whole and for future generations; (c) dealing with situations in which the economics of an activity being regulated are so thoroughly integrated nationally that there is no other way to regulate it (e.g., air travel, motor vehicles, interstate transport of hazardous wastes); (d) controlling interstate competition and granting all citizens a certain minimum level of health and welfare protection by having the national government set minimum, nationally uniform health and welfare standards (e.g., ambient air and water quality standards); and (e) establishing nationally uniform technology-based standards for a variety of reasons, including equity and fairness (e.g., among different plants in the same industry and among the different states), and ensuring that environmental quality "progress" continues by making sure new plants are cleaner than the ones they replace (Pedersen, 1982).

This structure of environmental and resource management also attempts to recognize differing perspectives of problems and issues (problem recognition and definition) that are related to scale and the appropriate level of governmental jurisdiction. More specifically, in many respects, the present system seems responsive to the focus at local scales on relatively short time frames and small spatial extent. However, at the same time, this system attempts to mobilize the potential of actors at the national level to focus on longer time frames (thus increasing the possibility of addressing transgenerational issues) and larger spatial extent.

There were several specific reasons, during the late 1960s and early- to mid-1970s (the period of greatest environmental legislative and regulatory activity), that Congress was concerned to have environmental policy making take on its present configuration (Strohbehn, 1982).[1] These reasons fall into two sets, the first of which is procedural and conceptual, and the second of which is substantive and pragmatic. In the first set, there are three main issues. The first is the issue of state sovereignty, which has two aspects. Traditionally, states have had primary responsibility for protecting health and welfare, and Congress was therefore concerned to have states play a key role in environmental matters. In addition, under the U.S. Constitution, the federal government cannot coerce the states to perform particular functions; therefore, Congress built into the environmental laws and regulations a number of inducements to encourage states to play the federally defined role.

A second conceptual/procedural issue concerns Congress's view of the states as laboratories where new ideas can be played out. This has resulted in some flexibility in the manner in which environmental policies are actually implemented; and this, in turn, has allowed for the resolution of some political and substantive problems within a federalist system (i.e., as a way to deal with the "one-size-fits-all" problem).

The third issue in this set has to do with several underlying concepts regarding governance. One concept, which for many years went largely unexamined, is the political science notion that decisions made by governmental entities closer to the people being governed are likely to be "better," since policy makers or politicians presumably are more accountable, and since their constituents have more access at smaller political scales. A second concept concerns efficiency and effectiveness; it postulates that decentralized policy making is more efficient and cost-effective because decisions can be made more quickly and require fewer resources. Of course, these notions of "better" and more "efficient" or "effective" depend largely on the interests being served; and certainly a number of areas have not been "better" served by local control (e.g., race relations, public education, health and welfare, etc.). Local control is also problematic in terms of environmental quality, because a state or local decision to improve environmental protection generally means, in economic terms, that the state or locality has imposed a price on what was otherwise a free good. This raises the competitiveness issue and leads directly into the second set of issues that were part of congressional deliberations when establishing the present system. It should also be noted here that many of the arguments being proffered for restructuring rehearse the critiques just cited.

From a substantive and pragmatic view, Congress was particularly concerned to prevent industrial arena-shopping and state and local competition to attract tax revenues. Since almost all of the resources for running state and local governments come from property and sales taxes or both, these entities have a built-in desire for attracting tax-base expanding development endeavors to their jurisdictions. As potential investors assess the various factors that make up locational desirability, highly varying environmental standards could play a large role. This is still a problem. For example, recently the Arkansas "Pollution Control and Ecology Commission lowered wastewater treatment standards for three Tyson Foods chicken-processing plants lest the company consider shutting down or moving the plants to other states with more favorable water-quality standards" (Rauber, 1995a, p. 27). Any move away from uniform national standards, therefore, would highlight spatial disparities and could result in a very different mapping of economic activities.

A second substantive/pragmatic issue facing Congress was to be responsive to the needs of the regulated community and to reduce the regulatory complexity facing industries. In this regard, a set of national objectives and standards was seen, almost universally, as preferable to objectives and standards being promulgated differentially by 50 state governments and thousands of localities. Again, restructuring would pose serious difficulties for both the regulated community and for environmental health and safety. However, monitoring and enforcement was another matter, with the regulated community preferring these activities to take place within the normal police powers of the state (and by authorization, local) governments. Under most current environmental laws, this model has been used. Goals, objectives, and standards (e.g., for clean water or air) have been set nationally, as have the means for meeting these goals. Implementation, monitoring, and enforcement activities have mostly been given over to states as they assume primacy under conditions set and overseen by the appropriate federal agency.

The final such issue facing Congress was to build into the environmental statutes, regulations, and programs an ability to deal with the fugitive nature of most environmental problems (i.e., most pollution concerns transcend political jurisdictions). In most cases, the relationship between a polluter and those affected is somewhat asymmetrical (i.e., the upstream/downstream problem). When such an asymmetry occurs between states, the conflict must be resolved by a higher authority, usually the federal government.[2]

In sum, at the time of the enactment of most environmental legislation, there seems to have been a consensus that the federal government should handle (a) interstate pollution problems; (b) issues in which significant economies of scale could be realized (e.g., training, research and development, etc.); and (c) areas in which national uniformity or preemption is necessary in order to avoid burdens on commerce (e.g., a need for uniform automobile emission standards), or to provide uniform, minimum health standards for all citizens, to avoid competition among states or localities, or to preempt parochial vetoes of projects

Environmental Policy and Restructuring 239

with important national interests (e.g., hazardous waste disposal, or the siting and monitoring of nuclear power facilities) (Smith, 1982).

This, in turn, provided some consensus on the proper role for state and local governments. First, it is important to note that the role was thought to be residual (i.e., whatever issues the federal government would not cover). Second, state and local governments were seen primarily as implementers and enforcers of national standards and programs. Third, state and local governments were viewed as important laboratories for regulatory and programmatic experimentation, taking on as much responsibility as possible to fulfill the operative belief that decision making should be decentralized for efficiency whenever practicable. Finally, almost all environmental legislation indicated that states have the right to be more stringent than the federal standards, except when preempted by overriding national interests (Smith, 1982).[3]

■ The Debate on Restructuring Environmental Management

As indicated in Kodras's chapter on the nature of state restructuring, the present round of restructuring generally takes one of three forms: *devolution* (moving services and activities to another level of government); *privatization* (shifting services and activities from the public sector); or *elimination* (the removal of services and activities). It seems to me important to think carefully about the differences (and connections) between these forms of restructuring. While devolution still acknowledges a useful and legitimate role for government services (but argues about the proper scale for delivery), privatization delegitimizes government's role altogether.[4] Devolution is also often a cynical mask in the sense that advocates know that frequently there is public resistance to outright privatization (e.g., of national parks), but moving governmental control to state and local levels accomplishes the same purposes in terms of resource access and exploitation, without a change in either ownership or the attendant responsibilities. Similarly, privatization can often be equated with program or service elimination, since advocates frequently anticipate that no private sector support will be found.

This differentiation provides some insight for spatializing this discussion a bit, at least in terms of environmental or resource issues. For example, on a macro-regional scale, restructuring the management of public lands and resources has very different implications in the eastern United States than in the West. Because in the West there is much more public land (602 out of approximately 630 million acres of federally-owned land; Lehmann, 1995, p. 22) and resources such as minerals, timber, grazing land, oil and gas (including 60 million acres on private land; Lehmann, 1995, p. 22) in the West, there is much greater interest in that part of the country in devolution, privatization, or marketization than there is in the East.[5]

Arguments for Devolution and Disappearance

These distinctions, of course, have been the basis for a number of arguments regarding the fundamental questions of the legitimate role of government. In natural resource and environmental issues, these debates have been most poignantly captured in the ongoing contest over such matters as the legitimacy of public welfare and public goods that pervades this discussion of the role of government. In shorthand terms, this is now embodied in the debate over so-called private property rights and the associated matter of "takings."[6]

At the heart of the private property dispute is what Associate Attorney General John R. Schmidt calls a "radical premise that has never been a part of our law or tradition: that a private-property owner has the absolute right to the greatest possible profit from that property, regardless of the consequences of the proposed use on other individuals or the public generally" (quoted in Jost, 1995, p. 529). But even critics like Schmidt pose the issue purely in economic terms: "First, we would have to give up on worker health and safety, environmental regulations, human health, and so forth, because passage of such legislation would be too costly if compensation were included; second, paying property owners to follow the law would also be extremely costly." In neither of these arguments is the notion of *public* property or *public* good advanced. This conception is rapidly being cut out of the discourse entirely.

To conform to contemporary rules of rhetoric, the necessary dualism has been constructed, and the debate has now given rise to two sets of "Bills of Rights." The first was embodied in a piece of legislation introduced in the first session of the 104th Congress as H.R. 790 "Private Property Owners' Bill of Rights" and was based on the following "findings" (among several others): (a) democracy is founded on principles of ownership, use, and control of private property (and these principles are embodied in the 5th Amendment to the Constitution that prohibits the taking of private property without just compensation); (b) several federal environmental programs (especially the Endangered Species Act and § 404 of the Federal Water Pollution Control Act) have been implemented in a manner that deprives private property owners of the use and control of their property; (c) private property owners have been forced into suits against the federal government to protect their basic civil rights; and (d) although all property owners should not use their property in a manner that harms their neighbors, these laws have traditionally been enacted, implemented, and enforced at the state and local levels where they are best able to protect the rights of all private property owners and local citizens. The purpose of H.R. 790 is to "provide a consistent Federal policy to encourage, support, and promote the private ownership of property and to ensure that the constitutional and legal rights of private property owners are protected by the Federal Government, its employees, agents, and representatives" (H.R. 790, § 2.b.).

According to Scott Lehmann (1995)—writing particularly about public lands, but whose arguments apply to resource and environmental issues more generally—there are several "general ethical conceptions that seem to provide a natural habitat for the privatization arguments: the appeal to productivity [which is the main appeal of privatization proponents] suggests a rule form of preference utilitarianism (private property maximizes the satisfaction of desires) while less-prominent appeals to individual freedom suggest a more Kantian outlook (private property secures the integrity of persons)" (p. 20).

As Lehmann (1995) argues, "The central appeal here [i.e., in the argument for privatization] is to productivity: public lands should be privatized because resources are utilized more productively when they are privately owned" (p. 17). But according to Lehmann, this appeal to productivity raises a number of provocative questions, including: (a) What, precisely, is meant by productivity? (i.e., how are competing preferences reconciled?) (b) Would the market in federal resources created by privatization really increase their productivity? (c) Assuming that it is good to put resources to their most productive use, whatever that means, is that the only good? (e.g., what about equity issues?) (d) What, precisely, is good about greater productivity? and (e) Is the attraction to more productive allocations of resources a matter of their history, the fact that they typically result from the autonomous actions of individuals in free exchange? (pp. 18-19).

These arguments, and others like them, have been made manifest in several kinds of concrete actions at various levels of government and by nongovernmental actors. For example, "[i]ncremental privatization [i.e., the conversion of de facto rights into de jure rights, by first recognizing such rights as private property, so that holders would have to be compensated] is a goal of the so-called County Movement, a successor to the Sagebrush Rebellion" (Lehmann, 1995, p. 11). The strategy of the County Movement has been to encourage county governments to enact land-use plans supportive of a ranching, logging, and mining culture and to direct federal land managers to respect them. An interim land-use plan in Catron County, New Mexico, for example, "stipulates that '[a]ll natural resource decisions shall be guided by the principles of protecting private property rights, protecting local custom and culture, maintaining traditional economic structures through self-determination, and opening new economic opportunities through reliance on free markets' and that '[f]ederal and state agencies shall comply with the county plan'" (Lehmann, 1995, p. 11). Thus far, 17 states have passed some form of "takings" legislation, although at present, most call only for a takings analysis before projects are undertaken or regulations adopted (Jost, 1995).

Another example of privatization strategies is the move to so-called free-market environmentalism. In a recent piece in the CATO Policy Report, Ridgeway (1996) discusses the merits and advantages of this strategy of privately protecting places and opposes it to command and control government systems. She visited a number of privately protected areas around the country—with grant support from the Political Economy Research Center and the Competitive Enterprise Institute, "two leading free-market environmental groups" (p. 10). She describes the solitude and undeveloped state of these privately protected lands and contrasts this with the crowding, roads, and so forth (i.e., the consequences of public access) encountered on the government-controlled lands.

Ridgeway (1996) goes on to indicate that "[o]ften owners [of privately protected lands] have no desire to open their property to visitors, because their concern is to minimize the damage too many visitors cause (witness Yellowstone or other state-owned parks)" and that "[m]any National Parks face the constant problem of preventing the great numbers of visitors from damaging what they have traveled so far to enjoy. Trail erosion, litter, and even intentional destruction and vandalism of natural treasures are all unavoidable problems on public property. That 'tragedy of the commons' does not occur on protected private property" (p. 13). After viewing these "privately protected places," Ridgeway concludes that "[w]e [the authors] learned much more [about free-market environmentalism] as we talked with, and worked alongside, individual landowners, the real environmentalists, who are motivated by love and respect for the land—their land" (p. 15).

Another alternative approach to governmental resource management and environmental protection is marketization (rather than privatization). Here the federal government retains ownership; but use of the lands, minerals, and other resources is controlled by market structures (e.g., auctions). For the most part, this strategy is seen by opponents as simply a politically expedient way to overcome the public's general antipathy to privatization. However, it accomplishes the same goals of easing access to, and control over, resources that are now seen at least nominally as "public" goods.

And, of course, the stakes are high. According to Lehmann (1995), "[T]he National forests contain half of the nation's softwood timber inventory. . . . In 1991, Federal energy leases contributed about 17% of the nation's domestic crude oil production, 31% of its natural gas, and 26% of its coal." In addition:

> USFS and BLM lands were estimated to be worth about $500 billion in 1983, about half the national debt at the time. After a decade of enormous deficits, their relative value has probably fallen to nearer a quarter of the debt. But we are still talking about a lot of money, and it is not out of the question that a comprehensive debt-reduction plan, should we ever settle on such a thing, will include liquidating some assets. . . . In the past, privatization advocates have certainly viewed the "debt crisis" as an opportunity to advance their cause. (p. 14)

And while the idea of outright privatization is still somewhat problematic, "the same cannot be said of some of the ideas behind it. In particular, the notion that resources are best allocated by a free market increasingly influences debate on Federal land policy" (Lehmann, 1995, p. 8). Furthermore,

> external factors will also help privatization advocates make their case. Socialism is discredited and under attack throughout the former Soviet empire. We will be invited to notice socialist structures within our own economy and to emulate our comrades to the East in dismantling them, and what, it will be suggested, could be a more obvious place to start than Federal lands [and other public resources]? (Lehmann, 1995, p. 13)

The Resistance

Of course such privatization and dismantling strategies have provoked a strong reaction within the environmental community. One long-standing counterargument points out that the debate is continuing to be contested entirely in economic terms. As characterized by Carl Pope of the Sierra Club, the privatization argument runs thus: "If the winners want clean air, wildlife, or recreation, the Social Darwinists argue, they can use the fruits of their competitive success to purchase them in the marketplace" (Pope, 1995, p. 14). From a slightly different perspective, Raena Honan (Legislative Director of the Grand Canyon Chapter of the Sierra Club) says, "Paying people to obey the law is called, where I come from, a 'protection racket'" (Honan, 1995, p. 5).

Is the environmental community right to be concerned, or are its members just paranoid? As former Nixon advisor John Ehrlichman once commented, "Even paranoids have real enemies sometimes!" Environmentalists, and others, are reacting to what they see as a new, concerted threat, embodied in the group of congressional representatives who came into office after the 1994 election, whose rallying cry was given form by the Contract with America, and whose numbers gave the Republican Party a majority in both houses. A few examples should suffice to give the tenor of the discourse coming from those now in charge of the U.S. Congress (or echoing similar ideas at other levels of government).

According to Barbara Cubin, a newly elected representative from Wyoming, "The Federal government doesn't have a right to own any land . . . except for post offices and armed forces bases" (quoted in Rauber, 1995a, p. 24). Gary Marble (a Republican state representative in Missouri), discussing the topic of "audit privilege" (polluter secrecy) legislation, offers the following assertion of why polluters should be able to keep their crimes a secret: "Because businesses and industries pay more taxes than common citizens, they should be granted more rights" (quoted in Rauber, 1995b, pp. 26-27).

One of the more powerful players in all of this is Don Young (R-Alaska), the new chair of the House Resources Committee—it used to be the House Natural Resources Committee, but the new leadership found the emphasis on nature offensive:

> Young (LCV [League of Conservation Voters] rating 0) replaced George Miller (D-Calif., LCV rating 92); he promises to use his reign as chair to

show people who environmentalists are, "the people who drive around in their limousines and live in their big mansions and say 'I'm an environmentalist'" and how elitist they really are. How elitist is that? They're the most despicable group of individuals I've ever been around . . . the self-centered bunch, the waffle-stomping, Harvard-graduating, intellectual bunch of idiots that don't understand that they're leading this country into environmental disaster. (Rauber, 1995a, pp. 24-25)

Another example of the perceived assault on the environment: In January of 1995, "the Appropriations Committee received with approval recommendations from Scott Hodge of the Heritage Foundation to halt funding for the Bureau of Land Management [BLM] and to abolish the National Biological Survey and even the venerable U.S. Geological Survey" (Rauber, 1995a, p. 26). This Heritage Foundation proposal would consolidate BLM, the U.S. Fish and Wildlife Service, the National Park Service, and the U.S. Forest Service into a single natural resources agency, and then have the agency turn back most of the lands it controls to the states over a 5- to 7-year period, keeping only the crown jewels (Yellowstone, Yosemite, and the Grand Canyon).

Finally, the environmental community has a continuing concern with the private property movement (and the related "takings" issue):

How far will the Fifth Amendment revisionists go? . . . Nearly as loopy [as a Texas takings group's suggestion that Lincoln should have paid for the slaves] is Oregon's proposed Future Ecotake Bill, which would require the state to compensate any landowner who did just about anything in the way of protecting, providing for, or preserving any aesthetic resource, including scenic areas, natural areas, open space, wildlife areas, wetlands, wilderness areas, outdoor recreation areas, and sites of historical, archaeological or cultural significance. (Rauber, 1995a, p. 29)

Given these concerns, the Sierra Club and others have deployed an "Environmental Bill of Rights" that affirms every American's right to "a safe, secure, and sustainable natural environment" (Bergman, 1995, p. 79) as the requisite rhetorical counter to such documents as the "Private Property Owners Bill of Rights." The Sierra Club and the Public Interest Research Group (PIRG), with the cooperation of Greenpeace, the Natural Resources Defense Council (NRDC), the Audubon Society, and others, have begun a petition drive aiming to collect more than a million signatures to present to elected officials in Washington, D.C. The petitions are designed "to wake up politicians, but they have another equally important objective: mobilizing grassroots opposition to polluters' efforts to gut a quarter century of environmental protections" (Bergman, 1995, p. 79). The Sierra Club has also launched an Environmental Rights Activist Network for members who want to take part in a campaign to stop what the club refers to as "the War on the Environment."

The environmental community has also begun to take advantage of the internet and the World Wide Web, with home pages for most of the major environmental organizations. For example, the NRDC has an online newsletter, as well as an ongoing series of reports with such titles as "Breach of Faith: How the Contract's Fine Print Undermines America's Environmental Success," and "The Year of Living Dangerously: Congress and the Environment in 1995." These mechanisms are used to provide environmental activists with up-to-date information on legislative activities, as well as links to others in the community who want to take action.

■ Restructuring Environmental Management Since 1994

The rhetorical lines have been drawn, and the contest has now turned to more material manifestations. Although in this volume we are not addressing the Contract with America per se, that document does represent an important rhetorical ploy and certainly articulates many of the ideological positions deployed in proposed and enacted legislation since 1994. It is also interesting to note that the Contract never mentions the word *environment*. However, two sections of the Contract have been used as the basis for what the Sierra Club terms a hidden "Polluter's Bill of Rights." Section 8 of the Contract, the Job Creation and Wage Enhancement Act, promises to enact "small business incentives, capital gains cuts and indexation, neutral cost recovery, risk assessment/cost-benefit analysis, strengthening the Regulatory Flexibility Act and unfunded mandate reform to create jobs and raise worker wages." Section 9, the Common Sense Legal Reform Act, promises to enact 'Loser pays' laws, reasonable limits on punitive damages, and reform of product liability laws to stem the endless tide of litigation."

Polls indicate that nearly 75% of those who actually voted in the 1994 elections had never heard of the Contract at the time. Before examining the specific environmental outcomes, it is important to ask, therefore, What were people really voting for in November 1994? According to opponents, since so few voters had heard of the Contract, and since it was likely that many would disagree with the particulars necessary for carrying out the provisions, even if they agreed with them in principle, there was little if any basis on which to claim a mandate from the November 1994 election. And since the word *environment* is never mentioned explicitly, opponents maintain that there is no evidence that people on election day in November 1994 were clamoring for dirty air and water, the loss of national parks, and so forth. Nevertheless, the Republicans have moved forward, since taking office, and have proposed or enacted a variety of legislation focused on natural resource and environmental issues.

Congressional Efforts to Date[7]

The proposed changes have come in the three main forms discussed above (and by Kodras) and are intended to either devolve functions to lower levels of government, shift them to the private sector, or delegitimize them entirely. First, the new majority has undertaken outright efforts to alter or eliminate existing laws (through devolution to other governmental levels or through privatization). These efforts have been mostly unsuccessful, given the continuing high levels of public support for environmental protection and governmental control of important resources and the ability of the environmental community to mobilize that support. Second, the Republican majority has deployed efforts to bog down the regulatory process in red tape and legal actions. These strategies have included the so-called regulatory reform and takings bills and have been underpinned by the rhetoric discussed above. These efforts have had mixed success in the legislative process, because these maneuvers can be packaged so as not to appear as direct assaults on publicly valued environmental and resource laws. Finally, Speaker Gingrich and his colleagues have attempted to alter environmental and natural resource management through the budget process. This has proven to be the most effective mechanism for the Republicans because it does not appear to alter the basic laws themselves. A brief synopsis of activities undertaken as of the summer of 1996, organized by environmental sector, will provide some insight into the nature of these changes.

Water

H.R. 961 (passed House)—This bill would relax standards for toxic releases, convert strong cleanup requirements for the Great Lakes into voluntary guidelines, ease requirements for polluted farm runoff, and weaken wetlands protection. Due to unexpectedly strong opposition, the companion bill has stalled in the Senate.

Regulatory Reform

H.R. 1022 (passed House)—This bill requires extensive, lengthy, and costly risk analyses and cost-benefit analyses before regulations can be promulgated; allows anyone to petition an agency to revise *existing* regulatory programs on the grounds that they do not comply with the cost-benefit and risk-assessment criteria of the bill; makes cost and risk criteria the overriding standard for health and safety regulation (as opposed to current programs, which use public health standards or technological effectiveness as the primary criteria for regulation); and allows those with conflicts of interest (including those at state and local levels) to sit on peer review panels for evaluating new regulations. No conclusive action has been taken in the Senate.

Takings

H.R. 925 (passed House)—This bill stipulates that landowners must be compensated whenever an agency, acting to save a wetland or an endangered species, limits property owners' use of land.

S. 605 (under consideration)—This bill, called the "Omnibus Property Rights Act," calls for payment when any federal protection program reduces the value of a property by a third.

Natural Resources

Proposals—A series of bills that would allow escalation of logging in old growth forests, the perpetuation of current grazing policies and fees, and the continuation of the 1872 Mining Act.

S. 391 (under consideration)—This bill would allow widespread logging in national forests. A previous budget rider (see more below), which passed the Congress and was signed into law by President Clinton in the summer of 1995, essentially allows "logging without laws" in the national forests. However, this policy is currently being reassessed.

S. 1054 (under consideration)—This bill would allow clearcutting in the Tongass National Forest in Alaska.

H.R. 2032 and S. 1031 (under consideration)—These bills would transfer all Bureau of Land Management lands to the states.

There have also been a variety of efforts to open the Alaskan Arctic National Wildlife Refuge to oil drilling.

Policy Riders and Funding Cuts

A variety of policy riders have been attached to budget bills, including provisions (among others) to block efforts to protect wetlands, open up the Arctic National Wildlife Refuge for oil drilling, undermine enforcement of air pollution requirements for key industries, allow the destruction of large segments of national forest, and "give away billions of dollars of America's natural resources to special interests" (NRDC, 1995).

So far, bills have proposed a 23% cut in overall Environmental Protection Agency (EPA) spending (measured against President Clinton's budget request); these proposals come on top of a $600 million funding recision imposed on the agency in July 1995.

Specific reductions in EPA programs include enforcement (22%); superfund waste site cleanups (26%), as well as a prohibition on listing any new sites; programs to address climate change (40%); and bans on listing any new species under the Endangered Species Act and on the protection of additional habitat to preserve any species already listed as threatened or endangered (included in a Department of Defense Supplemental budget bill).

There has also been a 50% reduction in the budget for the President's Council on Environmental Quality, the organization that oversees the implementation of the National Environmental Policy Act. Critics see this as a particularly cynical cut because it results in saving the federal government only $1 million.

There have also been a number of significant research funding cuts: climate change studies at the National Oceanic and Atmospheric Administration (36%); Department of Interior information collection on threatened or endangered species (complete ban); Department of Energy efforts to develop renewable energy supplies (35%) or increase energy efficiency (34%).

Finally, there have also been sizable reductions in the U.S. contribution to international research efforts, including a 33% overall reduction for such programs as the Montreal Protocol Fund (to reduce ozone-depleting chemicals), the World Meteorological Organization, the United Nations Environment Programme, and the International Union for the Conservation of Nature.

■ Conclusion

The Republican majority in Congress targeted environmental and natural resource issues heavily in their attempts to restructure and reorient government. Although a number of the more visible and overt battles have been won by environmentalists (through successful discursive strategies to mobilize the considerable public support for environmental protection), Congress has succeeded in making substantial changes through the budget process and other "back door" mechanisms. The constancy of public support for strong environmental protection (as evidenced by numerous opinion polls) has tempered the vigor with which the Congress has been able to dismantle, discredit, or delegitimize existing environmental legislation. However, as contests continue over public versus private property, budget and tax cutting, and the appropriate role of the state (at whatever level), the map of environmental and resource management will continue to be redrawn. By reconsidering the background to the current framework, it is possible to assess the ways in which restructuring might affect this mapping.

It is fitting that the most prominent ideological bumper-sticker statement relative to scale comes from the environmental arena as people are urged to "Think globally and act locally." This is an apt reminder that the fundamental issues of health, safety, and aesthetics with which all of us contend on a daily basis are an interesting mix of policy outcomes from the international to the local scale. We are witnessing the latest in a series of contestations over the appropriate locus of such activity.

NOTES

1. There were also very active debates over the ways in which environmental quality issues were to be integrated (or not) with natural resource management issues, but those debates are beyond the scope of the present paper. However, it is at least necessary to point out these two areas of concern are now managed independently of, and sometimes in conflict with, each other. Where appropriate, these conflicts will be noted in the discussion below.

2. This asymmetry can also be seen with respect to natural resource management, although in this case the asymmetry is frequently temporal (i.e., nonrenewable resources used today are unavailable for future generations). Again, however, appeal to a higher authority is usually required to resolve such matters and to speak for the public good.

3. An interesting set of cases in this regard came about as municipalities tried to prevent nuclear waste from being transported through their "nuclear-free" jurisdictions. It is also interesting to note the diminution of state and local control over environmental issues that accompanied the "fast-tracking" of the North American Free Trade Agreement.

4. Although, as I will discuss below, some more mixed restructuring strategies, such as marketization, are being advocated in many quarters.

5. The federal government owns the following proportions of land in the various western states: UT, 64; ID, 63; CA, 61; AK, 59; WY, 49; OR, 48; AZ 43; CO, 34; NM, 33; WA, 29; MT, 28 (Lehmann, 1995, p. 23).

6. Interestingly, environmentalists, and others, are now advancing a counterargument for the concept of "givings" (or goods and services provided by government) for which recipients should pay.

7. These materials have been derived from a number of sources, including the Natural Resources Defense Council's State of Nature newsletter and the Thomas on-line legislative tracking system. The enormous amount of legislative activity (including budget-related legislation) makes an accurate and up-to-date digest difficult to compile. However, the intent of this section is illustrative rather than comprehensive.

REFERENCES

Bergman, B. J. (1995). Standing up for the planet. *Sierra, 80*(3), 79.
Bruck, C. (1995). The politics of perception. *New Yorker, 71*(31), 50.
Honan, R. (1995). Capitol maul. *Canyon Echo, 31*(5), 5.
Jost, K. (1995). Property rights. *CQ Researcher, 5*(22), 513-533.
Lehmann, S. (1995). *Privatizing public lands.* New York: Oxford University Press.
Natural Resources Defense Council. (1995). *The year of living dangerously: Congress and the environment in 1995.* New York: Author.
Pedersen, W. F., Jr. (1982). Federal/state relations in the Clean Air Act, the Clean Water Act, and the Resource Conservation and Recovery Act: Does one pattern make sense? In American Bar Association, *The new federalism in environmental law: Taking a stand.* Washington, DC: Author.
Pope, C. (1995). Congress, red in tooth and claw. *Sierra, 80*(4), 14.
Rauber, P. (1995a). Stump speeches. *Sierra, 80*(3), 27.
Rauber, P. (1995b). Absolution for polluters. *Sierra, 80*(4), 26-27.
Ridgeway, C. (1996). Privately protected places. *CATO Policy Report, 18*(2), 1, 10-11, 13, 15.
Smith, T. T. (1982). Opening address: Reflections on federalism. In American Bar Association, *The new federalism in environmental law: Taking a stand.* Washington, DC: Author.
Strohbehn, E. L. (1982). The bases for federal/state relationships in environmental law. In American Bar Association, *The new federalism in environmental law: Taking a stand.* Washington, DC: Author.

14

Conclusion: Regional Collective Memories and the Ideology of State Restructuring

COLIN FLINT

As nation-states adapt to a globalized world economy at the end of the 20th century, balancing the imperatives of accumulation and legitimation (Habermas, 1973) is the key political dilemma. As Lake and Shelley argue in their chapters in this volume, the desire for social harmony calls for the construction of a polity that can simultaneously facilitate capitalist growth. The political agenda of the 104th Congress and the responses to it illustrate the tensions manifest in the need to preserve social peace in a context of aggressive global competition for capital investment. State restructuring is a process in which government, business, and citizens struggle over the form of political organization that will satisfy their competing demands for accumulation and legitimation.

As Shelley shows in his chapter, imperatives of social harmony and capitalist growth create tensions even within political viewpoints. In light of these contradictions, rhetoric is important in creating an atmosphere that facilitates radical state change (Kodras, Lake, Shelley, and Wright, all this volume). The goal of this rhetoric is to allow for changes in government structure that will reduce restrictions to the circulation of capital (Lake and Waterstone, both this volume) while justifying the burden faced by small businesses (Lake and Page, both this volume), labor (Herod, this volume), the poor and disenfranchised (Cope, Staeheli, Wolpert, and Wright, all this volume), and places that are economically marginalized by the global economy (Kodras and O'Loughlin, both this volume).

Collective Memories and Restructuring

As Kodras shows, globalization has allowed capital to abandon places that do not offer the desired combination of material and human resources while actively generating those that do. The resultant geography of opportunity within national borders emphasizes the important role of regions and places in state restructuring. Regions, defined by similar economic activities and cultural institutions, may be locations of either prosperous or obsolete economic activity at a given point in time. Within the broad pattern of regions, the particular social relations, political histories, and economic bases of local places create an even finer tapestry of opportunity and marginalization. For example, while the region of the South may be experiencing economic growth, a wide disparity of opportunity exists between metropolitan and nonmetropolitan places (see Kodras, this volume).

Policies that increase the mobility of capital, such as those overwhelmingly initiated by the 104th Congress, increase the importance of geography to investors. This one implication of state restructuring accentuates the geographies of marginalization and opportunity, as addressed by the contributions to this book. The policies themselves change the form of the state through the dismantling, devolution, and privatization of government activities to facilitate capital circulation. As government intervention in the process of accumulation is altered, place-specific attributes become essential in the competition for capital and in the legitimation of the system. Thus the current methods of state restructuring increase the burden on places to attract investment *and* provide a safety net for their citizens. This difficult balance is complicated by capital's ability to pick up and move if a more attractive combination appears elsewhere.

The purpose of this concluding chapter is to elevate the preceding investigations of state restructuring to a higher theoretical level, demonstrating the importance of geographical scale and place context to a full, conceptual understanding of recent governmental changes in the United States. I begin with a discussion of the political conflicts involved in state restructuring, as actors with a stake in the outcome attempt to alter the relationship of the state to capital and civil society, to reorient its dual role of supporting accumulation for capital and seeking legitimation from civil society. Specifically, I build on Schattschneider's (1960) early work on agenda setting to demonstrate that the process of renegotiating the state's agenda involves multiple dimensions of scale and place context. I show how each is an inherent part of the struggle to alter the balance in the state's dual role.

In the second section, I place the specific processes of state restructuring into the larger context of global political and economic transformations of the 20th century, based on Wallerstein's (1984) world-systems analysis. I use this framework to explain why each round of state restructuring occurs when it does and how it is linked to ongoing dynamics in the world system. Here I discuss both the New Deal period of expansion in the national state and the present period of retraction in the national state, demonstrating the shifting role of different scales of government activity over time. As the power of the national state recedes in the current era of economic globalization, regions and localities within states gain new importance, directly competing in the world economy. To conclude this section, I discuss how this interregional competition in the United States affects the politics of state restructuring. Specifically, each regional economy is sensitive to dynamics of the world system, rising and falling in concert with the match of its productive specializations to the shifting demands of the global economy. Those specializing in important new lead sectors are placed at an economic advantage, which can translate into political advantage as regions seek to maneuver for influence over federal policy. Powerful regions are best positioned to impose their own geographically specific and historically accumulated ideological perspectives on the federal government, restructuring the national ideological perspective on the nature and role of the state.

I further explore this notion in the third section, drawing on Gramsci's (1971) conceptualization of the processes involved in creating national ideologies. Whereas Gramsci focuses on the class politics implicated in this process, I emphasize the regional politics. I argue that interregional competition, regarded as a political imperative in the United States Constitution, requires an overarching national ideology that guards against the country's disintegration. Thus, to be successful in imposing their own ideological perspectives on the nature and role of the state, powerful regions must broadcast a vision of government that simultaneously promotes capitalist accumulation, advances the interests of the particular region, and is seen to benefit the country as a whole.

In the final section, I illustrate my argument using the example of the South, showing how one particular region is positioned to impose its vision of the appropriate scale and scope of the state. Although the region has been economically and politically marginalized throughout most of its history, the ascending South now possesses a regional tradition and ideology that resonates with the political perspective currently dominant at the national level. The popular expression of this tradition and ideology is a "collective memory"—the way in which a culture develops and is passed onto future generations (Johnston, 1991). Specifically, the ideological themes underlying the "collective memory" of the South include supporting states' rights over the power of the federal government, protecting private property rights, fostering "traditional family values," resisting civil rights and Affirmative Action, and incorporating practices of the Christian religion into public life. Although these themes are linked to the particular history of the South, they currently harmonize with concerns expressed elsewhere. The rhetoric surrounding calls for devolution, privatization, and dismantling of the national state finds substance and legitimacy in the traditions of the Confederate South and is given contemporary salience in an era of economic globalization.

■ The Politics of State Restructuring

The complexities and conflicts of state restructuring are underpinned by the process of agenda setting, the very essence of politics in the modern liberal democratic state (Schattschneider, 1960). Schattschneider argued that politics involves defining the extent of participation in a particular argument, with some actors believing that an increase in the number of participants would be to their benefit; whereas others perceive that the narrower the conflict, the more likely they are to succeed. From the geographical perspective, Schattschneider's idea is reflected in struggles over the spatial scope of a political conflict. Scope may be defined in two separate but related ways. The scope of a conflict may be defined by horizontal linkages across space as neighboring localities or states become involved. Alternatively, scope can be conceptualized as vertical linkages between a jurisdiction and higher scales of government activity in the federal hierarchy. Accordingly, I discuss below the role of geographic scale and place context in state restructuring.

As Mitchell argues in his chapter, the scales at which conflicts take place are an important part of politics. The dominance of the federal government in the period of the New Deal was of obvious benefit to particular groups—people of color, women, and labor especially (see Cope, Herod, Kodras, and Waterstone, all this volume). Federal courts, programs, and legislation greased the wheels of wealth creation, which benefited capital and the salaried middle class but also provided a degree of social mobility and legal protection for marginalized and more vulnerable sections of society. Those who wish to challenge these latter policies develop strategies that weaken the federal scale and consequently strengthen other scales. The calls for states' rights, various legacies of the Sagebrush Rebellion (see Waterstone, this volume), movements to increase the power of the county sheriff, militia movements, and so forth are all efforts to reorient the scale at which politics is undertaken. In contrast, an example of political strategies that try to widen the scope of the political conflict is seen in the effort of environmental and labor movements opposing NAFTA in particular and transnational corporations in general to create international support linkages as part of their political strategy (Rupert, 1995). Thus, one of the features of state restructuring is the definition of the scale at which political power will be located or, in other words, the scope of the political conflict. As people attempt to empower themselves, part of their strategy will be to control and strengthen particular scales of government activity and to weaken others.

Participation in politics is not just defined by geographic scale. Life chances in general, and political participation in particular, are also a function of the resources immediately available to people in the places where they live. As Kodras shows in her second chapter, the consideration of place context in the geographic perspective illuminates the economic and institutional factors that provide for spatial disparities in life chances. Social groups concentrated in localities with limited institutional and economic resources to further their political interests can be most easily marginalized by reducing the ability of federal authorities to act on their behalf (Wolpert, this volume). For example, the educational and human service orientation of suburban nonprofits, compared to the cultural and health focus of center-city nonprofits (Wolpert, this volume), increases the reliance of the central cities on publicly funded educational initiatives. The dismantling, privatization, and devolution of government activities gives the burden and initiative to places to provide a variety of services. Place-specific material and discursive resources define the consequent geography of marginalization and opportunity.

Now that I have summarized how scale and place are integral in the processes of state restructuring, I will discuss in more detail their role in balancing the twin imperatives of accumulation and legitimation. In the case of the United States, the political agenda has been dominated by pro-capitalist thought. The different scales of government have not operated as vehicles for an anti-capitalist message, with some exceptions in the early decades of the 20th century. Rather, scales of governance below the federal government have acted as venues for political arguments favoring strategies either of accumulation or legitimation, but within a pro-capitalist environment (see Kodras, this volume). The scale of government activity is the framework through which actors can attempt to widen or restrict the scope of the conflict over the particular balance of accumulation or legitimation functions.

At the outset of the New Deal regime, legitimation was an important consideration because unemployment and economic stress threatened the stability of the country. Federal control over wage policy was implemented to prevent pro-business state governments from reducing the level of wages in an attempt to attract investment. As the New Deal regime has come to an end within a context of the globalization of capital (see O'Loughlin, this volume), the politics of state restructuring involves a variety of efforts by state and local governments to facilitate accumulation.

Both the material and discursive elements of state restructuring currently in vogue advocate the locality as the most efficient scale for promoting accumulation. State governments are advertising localities within their jurisdictions as attractive sites for investment, and the role of the federal government as the guarantor of legitimation is being questioned. For example, early in the race for the 1996 Republican presidential nomination, Bob Dole was portrayed by the "people who knew how to get things done," such as the ex-Governor Lamar Alexander and businessman and capitalist tool Steve Forbes, as a dinosaur roaming the overbearing and obsolete halls of Washington. Later in his presidential campaign, "Citizen Bob" Dole championed states' rights (Gransbery, 1996) in order to distance himself from the federal apparatus. The federal government was viewed as an encumbrance to people's efforts to both accumulate capital and legitimate the system. Instead, local places were seen as the best venue to tackle these issues, with no discussion of how place context would reshape the geography of opportunity.

Actors who support the legitimation role of the state through such mechanisms as a social safety net undergirding the U.S. economy are very aware of the importance of defining the scope of a conflict via geographical scale. As Staeheli illustrates in her chapter, the local scale remains an important site of legitimation for citizens reacting to the decreased ability of the federal government to guarantee economic opportunity. However, the legitimating function does not necessarily occur within the local state itself. As globalization focuses the priorities of local government on the need for accumulation, citizens may feel abandoned by the local state; and in response, citizen groups throughout the country are building legitimating structures outside the arena of the state (see Staeheli and Wright, both this volume). Interaction between the federal government and both local government and citizen groups is made imperative by the realization that certain localities have neither the institutional nor the economic resources to provide their own social safety nets without federal assistance. The marginalization of certain places in the globalized economy requires them to appeal to a scale of government that is itself being marginalized in political restructuring (see Kodras, this volume).

Clearly, the efficacy of the federal government as a scale that can promote accumulation and legitimation is being challenged by processes of globalization. The ability of capital to move across state borders and the power of international finance over government decisions impinges on the role of the nation-state as a container and facilitator of economic activity. The result is "glocalization" (Swyngedouw, 1992) as capital is globalized; state functions are privatized, dismantled, or devolved; and labor remains localized. As the federal government loses some of its former saliency, its ability to garner revenue also declines, frustrating those who view the federal state as the provider of a legitimating safety net.

Thus, political conflicts over the role of government in fostering accumulation and legitimation in an era of global economic restructuring are permeated with issues of geographic scale and place context. The next section of this chapter discusses why the debate over the scope and scale of political activity is occurring at this particular time. I argue that broader systemic transformations in the global economy create a paradox in which globalization creates an increased importance for place and region. Rather than reducing the importance of geography, these broad transformations highlight the continuing importance of the geographic perspective taken in this book and the necessity of applying this perspective in the analysis of future rounds of political and economic restructuring.

■ Economic Restructuring

Wallerstein's (1984) world-systems analysis provides a general conceptual framework for understanding the process of state restructuring in the context of worldwide political and economic transformations over time. I use this framework to explain the changing role of the state in the United States within the larger dynamics of the world economy, from the New Deal era of the 1930s, when the federal state was granted great new powers relative to other scales, to the present era, when many of these recently acquired powers are threatened by devolution, privatization, and dismantling.

Specifically, Wallerstein (1984) traces the cyclical history of the global political economy in terms of 50-year Kondratieff waves, each consisting of a 25-year growth period (A-phase), followed by a 25-year period of stagnation and, significantly for our discussion, economic restructuring (B-phase). The creation of the New Deal regime and the challenges of the Contract with America and the 104th Congress to the type of state it had produced both fall within B-phases, periods of global economic restructuring during 1930-1940 and 1980-1990. Wallerstein (1984) identifies four aspects of economic restructuring in these B-phases that help to inform the nature of state changes that accompany economic restructuring:

> A) reduction of production costs of former core-like products by further mechanization and/or relocation of these activities in lower-wage zones;
>
> B) creation of new core-like activities ("innovation") which promise high initial rates of profit, thus encouraging new loci of investment;
>
> C) an intensified class struggle both within the core states and between groups located in different states such that there may occur at the end of the process some political redistribution of world surplus to workers in core zones; and
>
> D) expansion of the outer boundaries of the world-economy, thereby creating new pools of direct producers. (pp. 16-17)

The aspects of class struggle and economic restructuring (see Lake and O'Loughlin, both this volume) are of particular interest to an analysis of political and economic change within nation-states, rather than the global scale. The interaction of economic and political processes within Wallerstein's conceptual framework highlights the interrelationship between the accumulation and legitimation functions of the state. The two periods of state restructuring discussed (the New Deal and the contemporary changes) were efforts to achieve growth and social peace by balancing the economic and political imperatives defined by Wallerstein. The question remains whether the current round of state restructuring, within a context of globalization, can facilitate an "innovation" that renews economic growth and maintains the income and benefits of workers as state functions are dismantled, devolved, and privatized.

Resurgence of economic growth during a Kondratieff A-phase requires economic innovations that will act as a global lead sector. In the past, the country within which that innovation was developed had an initial advantage in the global economy that could be exploited to gain economic supremacy. The innovations typically consisted of a new product, combined with new production processes and social relations (Rupert, 1990). For example, the innovation bundle that led the past American century consisted of aerospace and electronic products, assembly-line production processes, and a social contract between management and workers that benefited both (Rupert, 1990). In other words, the economic relations of Fordism and the political regime of the New Deal were the innovational basis for the economic dominance of the United States during a previous upswing. The question remains whether current efforts at economic and state restructuring will help or hinder the position of the United States, and indeed its constituent regions, in the next upturn.

The cyclical pattern of the growth and decline in lead sectors of the global economy has, in the past, been reflected in the relative economic success of countries specializing in different productive sectors. However, recent structural changes in the operation of the world economy have altered the relationship between the site of economic innovation and the success of countries. "Glocalization" (Swyngedouw, 1992) has reduced the economic benefits accruing to governments from the success of private firms operating within their borders. As private corporations become increasingly free of ties to particular countries, other scales become important for capital. Now individual regions and localities with their specific "historical accumulation of assets and liabilities" (Agnew & Corbridge, 1995, p. 6) must compete directly within the global economy with less control by the national state itself. Regional expressions of common economic interests in the United States have increased in significance as regions assert their independence from the federal government, and the numerous advertisements in business magazines extolling the virtues of particular states and cities to potential investors illustrate the role of localities as agents of accumulation.

The dynamism of the world economy, conceptualized as Kondratieff waves, alters the role of regions over time. Those regions hosting lead sectors of the world economy will rise and fall with the importance of that activity (Kodras and O'Loughlin, both this volume). In large and diverse countries such as the United States, inter-regional economic competition promotes intra-federal state political maneuvering for influence over federal policies (Page and Waterstone, both this volume). Historically, this has had obvious implications for trade policy. However, in the contemporary context of globalization and an orthodoxy of free trade, trade policy has become less important; and the focus has turned toward issues of accumulation and legitimation within nation-states.

The regions of the United States are gaining saliency in both academic and semipopular work. The Borderlands have been the focus of much attention in the wake of NAFTA (see Wright, this volume), and the distinctiveness of the regional mosaic of the United States is captured by the title of Garreau's *The Nine Nations of North America* (1981). Regions have always played an important role in the electoral politics of the United States (Archer & Taylor, 1981), but their political influence is also ideological.

In the current period of restructuring in the United States, political rhetoric promotes a new regional ideology within which the dramatic changes taking place may be located and given meaning. A new hegemonic ideology will be an integral feature of state restructuring, as defined by the "collective memory" of the regions leading that change. The nature of national political rhetoric will reflect the essence of the contemporary hegemonic ideology, a message that must promote the goals of a particular region or regions without acting as a national centrifugal force.

By placing the rhetoric of political restructuring within the dynamics of the global economy, I have discussed the changing role of the state in an era of globalization. One of the implications of globalization of capital is the increasing importance of regions and localities as sites of investment, and their relative autonomy from the national state. Federal policies are still of importance to regions, however, requiring competing regions and localities to lobby the federal government. Indeed, the current round of state restructuring will increase the material and ideological significance of regions and localities, even as it sets regions against each other as they compete for capital investment through economic development and social welfare strategies. In the next section, I use a Gramscian approach to illustrate the discursive strategies that regions adopt to influence the federal government to their benefit.

■ Gramsci's Hegemony

Gramsci (1971) argues that the interests of dominant and subordinate classes within a nation-state are coordinated through a bourgeois hegemonic ideology in a way that promotes the goal of economic growth while preventing class conflict (p. 13). Historical blocs are formed that are temporally specific (and, I will argue, regionally specific) combinations of the ruling classes and subordinate groups. The politics involved in creating historical blocs is a dynamic process, and the particular nature of these can be related to structural imperatives at the global scale. However, the political conflict manifest in the creation of these historical blocs is not just national inter-class conflict, but inter-regional, intra-class competition as well. By focusing on the regional conflicts within the formation of Gramscian historical blocs, some of the specifics of U.S. state restructuring may be seen.

Gramsci (1971) noted that "international relations intertwine with ... internal relations of nation-states, creating new unique and historically concrete combinations" and that "this relation between international forces and national forces is further complicated by the existence within every State of several structurally diverse territorial sectors, with diverse relations of forces at all levels" (p. 182). The geographer's contribution to this analysis is a focus on regional conflicts as manifestations of state restructuring within an international context and as vehicles for political mobilization. As Kodras, Lake, O'Loughlin, and Page each show in their contributions to this book, regional disparities in economic well-being are a significant feature of contemporary economic restructuring. Individual regions advance different agendas, which are suited to their economic needs, into the national political arena. Regional disputes within countries result in a particular form of the national state related to the relative dominance of particular regions. The ideology and interests of dominant regions are translated into state power that is then resisted, to varying degrees, by other regional interests.

Gramsci's particular approach highlights the confrontational and class nature of regional political conflicts. Though the hegemonic ideology benefits the bourgeoisie as a whole by excluding alternative forms of economic and political organization from the agenda, inter-regional bourgeois competition, reflecting the dynamics of the world economy, will also exist. Both of these aspects of hegemonic formation are addressed in the American polity. In Paper Number Ten, the Federalist Papers purposely proposed a political system containing a variety of regions that would produce an internal balance of power (Fairfield, 1981). Inter-regional conflict was, therefore, seen as an integral and stabilizing element of American governance. However, such conflict was to occur between "men of property" to prioritize the accumulation of capital over redistributive functions that would help legitimate the system throughout the population.

Thus, Gramsci's theoretical framework provides a basis for interpreting the Federalist Papers and the regional geography of the United States whereby political struggles over the form and ideas of the federal government partially emanate from regional interests. Furthermore, the relative influence of particular regions in this discourse is linked to the dynamics of the world economy, which gives greater or lesser advantage based on regional productive specialization. Popular attitudes across the nation regarding the appropriate regime of accumulation and the proper means of legitimation will be based on regional discourses formed within the changing imperatives of the world economy. The following section illustrates how one region, the South, has struggled in the past to impose its own ideology on the national political agenda, and how that ideology currently resonates with national political attitudes.

■ Ideology, Hegemony, and the Assertive South

The South bears the burden of a unique regional history, for it lost any chance of forming its own state and was largely unable to prevent its institutions, ideologies, and practices from being challenged, eliminated, or restructured by dominant interests of the federal government at many points over time. In the current period of restructuring, by contrast, the South possesses a regional tradition and ideology that resonates with national political imperatives. In other words, the ideology that dominated in the South, but was long suppressed at the national level by political initiatives such as the New Deal, is currently presented as the ideology that best reflects interests for "the American people" as a whole, and that provides the best framework for a new type of state. The political rhetoric of current state restructuring as found, for example, in the *Contract with America,* finds substance and legitimacy in the traditions of the Confederate South and contemporary saliency because of globalization.

In response to the establishment of the New Deal regime, business leaders in the South formed and disseminated two related themes: the South was burdened by a "colonial economy" and Roosevelt's policies were an attempt to hinder southern economic growth and independence (Mertz, 1978). The South's "colonial economy" was evident from its specialization in low-value raw materials and semi-processed goods that were then finished and given added-value in the North (Mertz, 1978, p. 222). Wage increases decreed by the NRA (National Recovery Act) were the focus of southern business leader's attacks on the federal government under Roosevelt. Specifically, the Southern States Industrial Council (SSIC) was formed in 1933 to oppose federally imposed wage increases and to publicize the industrial potential of the region Mertz, 1978, p. 223).

The geographical element of the ensuing rhetorical battle between the New Dealers and the SSIC and its allies juxtaposed the notion of a progressive and united nation versus a distinctive and historically burdened region that would improve its lot if left alone. President Roosevelt's statement that "the South presents right now the Nation's Number One economic problem—the Nation's problem, not merely the South's" was clearly designed to place the South within a larger national context. In the South, however, Roosevelt's statement was seen as further evidence of the intrusion of national interests into the region. In 1938, the *Atlanta Constitution* wrote that Roosevelt's statement "was not so much a criticism of the South [as of] . . . short-sighted interests in other sections which have been chiefly responsible for the condition, to the extent that it exists" (quoted in Mertz, 1978, p. 236). In addition, George C. Biggers, the business manager at the *Atlanta Journal,* argued that the South was "the nation's number one economic opportunity" and that national policies were barriers to the region's growth (Mertz, 1978, p. 238). Finally, C. C. Gilbert of the SSIC accused northern manufacturing interests of attempting to fend off southern competition by pressing the federal government for a minimum wage (Mertz, 1978, p. 228).

Though many business leaders were vocal against the federal policies of the New Deal, the Great Depression demonstrated to municipal leaders that their own resources and a laissez-faire policy were inadequate (Gelfand, 1975, cited in Smith, 1988, p. 5). Specifically, the dual roles of the state—legitimation and accumulation—produced conflicting views over the value of the New Deal within the southern elite. Municipal leaders, who had greater interest in the legitimating role of the state, were more susceptible to the New Deal ideology than were businessmen whose comparative advantage was based on cheap labor. As Lake argues in this volume, the ideology of the New Deal was suited to a national economy facing economic recession and restructuring and was therefore able to appeal, to some extent, to all regions of the country sharing in tough times. Resistance to the dominant political ideology was evident in the South, but (as must be the case according to Gramsci) it did benefit the nation as a whole and was adopted. The economic imperatives of the Great Depression helped create a national hegemony that challenged the pro-accumulation ideals and institutions of the Old South and its paternalistic attitude towards legitimation.

The New Deal ushered in an economic regime with a particular ideology—one that favored federal government spending and intervention, emphasized the country as the economic unit, and reflected the importance of states in the world economy. However, as has been made clear in this book, both the economic and political contexts are currently changing. The New Deal regime has ended (Fraser & Gerstle, 1989; Lake, this volume) in concert with the changing form of the world economy (Agnew & Corbridge, 1995; O'Loughlin, this volume). Because the world economy no longer favors nation-states as the prime scale of political organization (O'Loughlin, this volume; Rosecrance, 1996), individual regions are reasserting their identity and economic power and proclaiming their own ideology to best suit their particular situation and goals. As O'Loughlin shows in this volume, the U.S. economy is best understood as a number of metropolitan and local labor markets, rather than as a single market and production point. This geography of diverse economic opportunities requires an ideology more favorable to regional independence and place-specific ideology than was the case for the New Deal.

The ideology underlying the "collective memory" of the South is uniquely suited to become the new national ideology. As discussed throughout this volume, the policy changes currently underway all reflect political and economic perspectives that historically have enjoyed the most fervent support in the South. More specifically, the rhetoric of the 104th Congress is an update of "Southern obeisance to state's rights and decentralized authority [that] had long resulted in a fear of federal government intervention" (Biles, 1994, p. 157).

The word *update* is crucial here. In 1949, V. O. Key Jr. saw that in-migration and industrialization were forcing change in the political organization of the South, but he predicted that the depth of this change would be limited: "These changes are altering the shape of the mold that influences, it does not fix, the shape of southern politics" (Key, 1949/1984, p. 672). From the geographer's perspective, the economic and political shifts that Key identified are grounded in a region's "collective memory," the sensibility through which people evaluate problems and formulate solutions. At the regional scale, "collective memory" underlies dramatic shifts in electoral behavior as well as attitudes toward government.

The updating of southern political ideals involves internal changes, such as industrialization and urbanization, as well as external influences, such as Yankee in-migration, but also the legacy of resistance to the material and discursive elements of the New Deal regime. States' rights and individualism may appear paramount in the political rhetoric of the South, but a pragmatic attitude to government intervention is also evident. This new Southern political attitude has been captured by Earl and Merle Black:

> The reigning political philosophy of the new southern middle class is the entrepreneurial version of the individualistic political culture, a blend of conservative and progressive themes. In its emphasis on low rates of taxation, minimal regulation of business, and resolute opposition to unions and redistributive welfare programs for have-nots and have-littles, the current political ideology retains important continuities with the traditionalistic political culture. Its progressive element consists in its willingness to use governmental resources to construct the public infrastructure—highways, airports, harbors, colleges and universities, research parks, health complexes—that in turn stimulates and makes possible additional economic growth. (Black & Black, 1987, p. 297)

It is these attitudes, adhered to and disseminated most strongly by the southern middle class, that are defining the contemporary political agenda at the national scale. The elements of this contemporary political attitude that elected the freshmen Republicans of the 104th Congress are generally couched in the neutral, or even exemplary, terms *entrepreneurial* and *pragmatic*. However, the underlying motives are pro-accumulation policies and the allocation of power to scales lower than the federal government. This "pragmatism" resonates through the "collective memory" of the South, as its roots are firmly established in the states' rights rhetoric of the Old South. In addition, the themes of morality and responsibility, identified by Staeheli as current popular concerns, are part of an idealized image of the Old South. Doubts about how inclusive this community will be are also reflective of the ideology of the Old South (see Shelley, Staeheli, and Wright, all this volume).

The vitality of the South's "collective memory" is partially a legacy of the Confederacy. Organizations such as the Southern League, based on the campus of the University of Alabama, actively promote a sanitized and favorable view of the Old South. The following is a statement of the Southern League's position taken from their web page:

> If the South were its own nation, its GNP would rank in the top five nations of the world. Its laws would better reflect the natural conservatism and Christian roots of the Southern people. We could enjoy low taxes, sound money, secure private property rights, and a free-market economy. We could follow a foreign policy of armed neutrality, leave the UN, and oppose the New World Order. We could once again reward merit and abolish the Welfare State and Affirmative Action. We could severely limit immigration. We could get government out of our children's education. We could remove ourselves from the current judicial tyranny. In short, we could seize control of our destiny as a distinct people. (Southern League, 1996)

The program of the Southern League illustrates how regional history and institutions can be selectively used to make claims for a nationwide ideology. The Southern League's manifesto is an attempt to reverse the New Deal regime which, from their perspective, impinged on southern values and practices. The extremist voice of the Southern League promotes an ideological agenda that is diluted by the majority of the electorate who combine it with the message of the New Deal regime and its greater emphasis on legitimation at the national scale.

In the current period of state restructuring, and more specifically state devolution, the political values of the South are being touted as applicable to the rest of the country. For example, the heritage of the

South emphasizes the now popular political themes of private property rights, anti-Affirmative Action, "traditional family values," and the incorporation of the Christian religion in public life. These themes are linked to the particular history of the South, but they resonate across other parts of the country. The promotion of private property rights is seen in conflicts over logging and grazing rights in the West, fears underlying challenges to Affirmative Action are seen in the arguments over access to public goods in the Mexican-border region, and the nationwide growth of religious right movements (such as the Promise Keepers) is a product of the increased visibility of Christian values in social and political debates. In response to these political initiatives with roots in southern ideology, the legacy of the New Deal ideology is producing voices concerned about cuts to Medicare and the removal of government supervision of certain industries—air travel and pharmaceuticals, for example. In other words, the pro-accumulation rhetoric of the South is partially ameliorated by pro-legitimation concerns stemming from the New Deal regime to form a new national ideology.

Thus, we have seen how the political themes of contemporary state restructuring emanate from traditions of the South but must also include attitudes formed during the New Deal regime. Those in the electorate susceptible to the idea of ending "big government" (Lake, this volume) may ground such restructuring in the common-sense appeals of a regional history that has had its fair share of federal government intervention. On the other hand, the Democrats' labeling of Republicans as "extremists" in the 1996 presidential campaign indicates that the electorate is still cognizant of the legitimating role of the federal state that was an important element of the New Deal regime.

The shifting patterns of political ideology witnessed in the current period of state restructuring have several geographical implications. The most obvious is the rise and fall of regions in terms of their role in formulating political agendas (Kodras, Lake, and O'Loughlin, all this volume). Also of importance is the ability of different localities to adjust to changes in the state that favor accumulation over legitimation (Cope, Kodras, Page, Staeheli, and Wolpert, all this volume). These changes are also manifest in the changing geographic scale of the state, in particular the decreased saliency of the federal scale. Finally, the electoral politics of the United States may further fragment as coalitions of interest groups within the two major parties are forced to address issues at the local scale at a time when the competition between localities is likely to increase within the demands of globalization.

■ Conclusion

To summarize my points, regional competition within nation-states is the manifestation of the intersection of international and national forces that Gramsci saw as a crucial element of intra-state political dynamics. The politics of state restructuring includes conflict over the scope or scale of political activity as a means of influencing the agenda and the outcome of that activity. Competing regions will use rhetorical arguments to define the scope and scale of political activity to their benefit. In the contemporary case, the South is positioning itself as the source of values that, according to their proponents, have a regional grounding and a national appeal.

Through an examination of the sources and nature of state restructuring and its implications across sectors and social groups, the need for a legitimating ideology has been eluded to. As once-accepted government functions are either transferred to the private sector, placed in the hands of local government, or lost completely, expectations of economic growth must be fueled and fears of impoverishment placated. To this effect, the legitimating role of the state is challenged as a handicap to wealth creation and social mobility, and the locality is promoted as the most efficacious scale of accumulation. Globalization requires that localities must compete across the globe to attract capital, while at the same time the dismantling, privatization, and devolution of state functions imposes the burden of legitimation on that very same scale that must be "pro-business."

A consistent theme throughout the contributions to this book has been the benefits accruing to big capital in the current round of state restructuring and the increasing vulnerability of marginal groups and places. As the federal scale loses its desire and ability to act as provider, legitimation of the system will increasingly have to be undertaken through discursive channels as material safety nets are scaled back. The three-pronged strategy of devolution, privatization, and dismantling will benefit those places with both the adequate material foundations and a suitably oriented civil society to foster both economic growth and social cohesion. Those places that can integrate themselves into the global economy will be able to proclaim the success of "common sense" and "pragmatism," whereas less fortunate locations must ponder on the means of building social capital when capital has passed them by.

The geography of opportunity and disadvantage elucidated in this book calls for an active civil society, aware of both the material changes that are occurring and the discursive ideology that justifies them. Most importantly, the roles of geographic scale and interaction between places in defining relative well-being require organizational capabilities that can traverse vertical and horizontal distances. A democratic society is one that facilitates inclusion, participation, and communication. Civil society can foster these goals by calling for the renewal of a legitimating state via the generation of a message that reflects the needs of all places and communities.

REFERENCES

Agnew, J., & Corbridge, S. (1995). *Mastering space: Hegemony, territory, and international political economy.* London: Routledge & Kegan Paul.
Archer, J. C., & Taylor, P. J. (1981). *Section and party: A political geography of American presidential elections from Andrew Jackson to Ronald Reagan.* New York: Wiley.
Biles, R. (1994). *The South and the New Deal.* Lexington: University of Kentucky Press.
Black, E., & Black, M. (1987). *Politics and society in the South.* Cambridge, MA: Harvard University Press.
Fairfield, R. P. (Ed.). (1981). *The Federalist Papers: A collection of essays written in support of the Constitution of the United States.* Baltimore, MD: John Hopkins University Press.
Fraser, S., & Gerstle, G. (Eds.). (1989). *The rise and fall of the New Deal order, 1930-1980.* Princeton, NJ: Princeton University Press.
Garreau, J. (1981). *The nine nations of North America.* Boston: Houghton Mifflin.
Gelfand, M. (1975). *A nation of cities: The federal government and urban America, 1933-1965.* New York: Oxford University Press.
Gramsci, A. (1971). *Selections from the prison notebooks* (Q. Hoare & G. N. Smith, Ed. and Trans.). New York: International Publishers.
Gransbery, J. (1996, July 29). Dole backs state's rights. *Billings Gazette,* p. A-1.
Habermas, J. (1973). What does a crisis mean today? Legitimation problems in late capitalism. *Social Research, 40,* 39-46.
Johnston, R. J. (1991). *A question of place.* Oxford, UK: Basil Blackwell.
Key, V. O., Jr. (1984). *Southern politics in state and nation.* Knoxville: University of Tennessee Press (Originally published by Alfred A. Knopf, 1949).
Mertz, P. E. (1978). *New Deal policy and southern rural poverty.* Baton Rouge: Louisiana State University Press.
Rosecrance, R. (1996). The rise of the virtual state. *Foreign Affairs, 75,* 45-61.
Rupert, M. E. (1990). Producing hegemony: State/society relations and the politics of productivity in the United States. *International Studies Quarterly, 34,* 427-456.
Rupert, M. E. (1995). (Re)politicizing the global economy: Liberal common sense and ideological struggle in the U.S. NAFTA debate. *Review of International Political Economy, 2,* 658-692.
Schattschneider, E. E. (1960). *The semi-sovereign people.* Hinsdale, IL: Dryden.
Smith, D. L. (1988). *The New Deal in the urban South.* Baton Rouge: Louisiana State University Press.

Southern League. (1996). *The Southern League home page.* http://www.dixienet.org.
Swyngedouw, E. A. (1992). The Mammon quest. "Glocalisation," interspatial competition, and the monetary order: The construction of new scales. In M. Dunford & G. Kafkalas (Eds.), *Cities and regions in the new Europe: The global-local interplay and spatial development strategies* (pp. 39-67). London: Belhaven Press.
Wallerstein, I. (1984). *Politics of the world-economy.* Cambridge, UK: Cambridge University Press.

Index

AAFRC Trust for Philanthropy, 105
Accumulation role of state, xxiii, xxiv, 6, 252, 256, 258, 269
ACF, 187, 193, 194, 195
Acreage Reserve Program (ARP), 143
Administration for Children and Families (ACF), 194-195
AFDC, 56, 192-193, 195
  geographical variation of benefits, 193-194, 198-199
AFDC-UP, 194
AFL-CIO, 179
Age Discrimination in Employment Act of 1967, 164
Agnew, J., 41, 91, 208, 210, 211, 260, 265
Agnew, J. A., 22, 33
Agricultural Adjustment Act of 1949, 148
Agricultural policy, restructuring of, 141
Agricultural Trade Development Act of 1954, 145
Alexander, Lamar, 257
American Association of Geographers (AAG), xi
Americans with Disabilities Act (1990), 164
Anderson, B., 210, 216, 217
Appaduri, A., 214
Archer, J. C., 261
Associational life, loss of, 63
Auster, L., 65
Authoritarian populism, 136

Bachrach, P., 8
Balderrama, F. E., 212
Ballenger, Cass, 174, 175
Baratz, M., 8
Barnes, T., 219
Barnet, R., 36
Barrow, C., 6
Basch, L. G., 211, 218
Bellah, R., 66
Bellanger, S., 12
Bennett, R., 83

Bennett, William, 230
Bergman, B. J., 245, 246
Bernstein, J., 10, 16, 26
Beynon, H., 90
Bhagwati, J., 29
Biggers, George C., 264
Biles, R., 266
Black, E., 266
Black, M., 266
Blakley, E., 43
Blaming government, xxiv
Block, F., 183, 184, 201
Blomley, N., 127, 130, 136, 137
Bluestone, B., 24, 27, 28, 29, 32, 36
Bonanno, A., 157
Borjas, G., 45
Borjas, G. J., 28
Bourdieu, P., 126
Bourne, R., 212
Bowen, W. G., 104
Bowers v. Hardwick, 120, 125, 137
Bowles, S., 8
Bowman, A., 87
Bratsberg, B., 209
Breuilly, J., 210, 213
Brown, Ron, 167
Brownback, S., 64
Brown v. Board of Education, 125, 225, 228
Bruck, C., 234
Buchanan, Patrick, xxix, 38, 136, 208, 213
  anti-abortion stance of, 223
  anti-government populism of, 16
  anti-immigrant sentiment of, 208, 223
  GATT rejection by, 223
  NAFTA rejection by, 223
  presidential campaign, 118
Bureau of Indian Affairs, 55
Bureau of Land Management, 245, 249
Bush, George, 128, 222, 228
Bush administration, 195
  education reform and, 229
Business unionism, 134

Index

Capital, xxiii, xxxiii, xxxiv, 89-90, 131
  globalization of, 162
  state and, xxiv
Capitalism, cultural diversity and, 223
Capital mobility, accelerated, 253
  barriers to, 13
  forms of, 11
  See also Geographical mobility; Organizational mobility; Sectoral mobility
Carnoy, M., 6
Carter administration:
  Program for Better Jobs and Income, 194
Castells, M., 28, 33
Catholic Charities, 99, 112, 116
Cavenagh, J., 36
Charities and charitable agencies:
  Americans' contributions to, 105-107
  consequences of expanded role of, 109-110
  description of, 100-101
  federal government and, 114-115
  impacts of federal cuts on, 110-112
  limits to increasing donations/voluntarism, 112-113
  negatives of reliance on, 97
  suggestions for, 113-114
  See also Nonprofit sector
Charles River Bridge v. Warren Bridge, 131, 137
Christian Coalition, 65
Chronicle of Philanthropy, 113
Chubb, J., 229
Cinel, D., 212
Cisneros, H., 65
Citizen's Project (Colorado Springs), 71
Civil Rights Act of 1964, 164
Civil Rights movement, 119, 126, 164
Civil society, xxiii-xxiv, xxxiii, xxxiv, 90
  morality and, 66
  virtue and, 66
Clark, G., xxii, 7, 90
Clark, G. L., 34
Clark, W. A. V., 213
Clinton, Bill (President), xix, 87, 128, 141, 161, 166, 169, 187, 195, 196, 198, 200, 202, 248
  1996 State of the Union Address, 61
  welfare criteria, 187, 199-200
Clinton administration, 3, 158, 161, 195
  education reform and, 229
  restructuring welfare and, 192, 196

Clinton, H., 66
Clotfelter, C. T., 102
Cloward, R., 183, 184, 187, 192, 201
Cochrane, W. W., 143, 151, 154
Cockburn, C., 83
Collective bargaining, local, 165
Colorado gay rights initiative, xxxi
Commins, P., 157
Commission on the Future of Worker-Management Relations. *See* Dunlop Commission
Commitment, second language of, 66
  responsibility in, 66-67
Commodity Credit Corporation (CCC), 145
  nonrecourse loans and, 143
Common school movement, 224, 229
  moral education and, 224
  political education and, 224
Common schools:
  curricula, 230
  ideology taught in, 224
Common Sense Legal Reform Act, 246
Communitarian Movement, 61, 67
Community:
  exclusion from citizenship and, 72
  revived importance of, 67-70
  roles of in citizenship, 68-69
  unity and, 67, 71
  volunteering and, 69
Conant, J., 225
Congressional elections of 1994, xiii, xvii, 86, 127
  campaigns, xxix
Connolly, W. E., 214
Conservation Reserve Program (CRP), retention of in FAIR Act, 146
Constance, D., 156, 157
Contingent workers, 176
Contract with America, ix, xiii, xiv, xviii, xxi, xxviii, xxxiii, 3, 13, 57, 87, 141-142, 222, 233, 246, 259, 264
  charity and, 97
  environment and, 233-234, 244
  "Personal Responsibility Act" and, 67
  policies and, xvii
Contract With America Advancement Act, Title II of, 175
Cooper, M., 133, 134, 135
Cope, Meghan, 123
Corbridge, S., 41, 260, 265
Corporatists, 83
  critics of, 83-84

Cox, K., 14, 89, 90, 91, 92
Creative destruction, 26
Cross-scale alliances, xxxi, 93
Cuban, L., 225, 226
Cubberley, E., 224
Cubin, Barbara, 244
Cultural diversity:
   capitalism and, 223
   government regulation of moral issues and, 223
Culture of rights, 136
   reaction against, 63-65
   See also Rights
Cushman, J., 4
Cushman, J. H., 151, 158
Cypher, J., 6

Darden, H., 45
Dartmouth College v. Woodward, 131, 137
David, P., 25, 26
Dear, M., xxii, 7, 90
De Certeau, M., 126
Dehejia, V. H., 29
Delayering, 26
DeMelo, J., 37
Depression, Great, xxxiii, 86, 98, 265
Deregulation, 167
Deskins, D., 28, 33
Detroit, economic restructuring and, 52-53
de Vita, C., 83
Devolution, ix, xiii, xxviii, 80, 83, 87, 89, 90, 196, 253, 256, 269
   as spatial process of change, 81
   definition of, 81
   of workplace regulation, 162, 173, 174, 175
   welfare and, 184
Devolutionists, 83
di Leonardi, M., 212
DiIulio, J., 84, 85, 94
DiNitto, D., 192
Dismantling, xxviii, 80, 83, 196, 253, 256, 269
   definition of, 82
   regulations and regulatory bodies, 162, 173, 174
   welfare and, 184
Dixon, S. R., 32, 34
Dole, Bob, 198, 223, 257
   education reform and, 229

Downsizing, 26
Duffy, E. A., 104
Dunlop, John, 167
Dunlop Commission, xix, 161, 162, 167, 168-173, 179
   employment law and, 171
   private alternative dispute resolution (ADR) and, 171-172
   recommendations, 168-169
Dye, T., 192

Economic globalization, x, 29-32, 36
   factor price equalization theory and, 30
   trade and, 30
Economic growth, negative, 17
Economic restructuring, 259-262
   political response and, 10-16
Education:
   as national problem, 225
   moral, 224
   political, 224
   reform of public, 224-225
   Republican dilemma and, 224-225
   See also Common school movement; Common schools; Education policy; Education reforms
Education policy:
   dismantling government bureaucracy and, 226
   Republican dilemma and, 226-231
Education reforms:
   financing education, 226
   improving competitiveness of students in U.S. schools, 226-227
   increasing local control, 226, 227-228
   school choice, 226, 228-230
Ehrenreich, B., 62, 63, 65, 73, 183, 184, 201
Eisenhower, Dwight D., 222
Elliott, M., 23
Ellis, M., 210
Ellwood, D., 46, 48, 56
Elshtain, J., 63, 64, 65, 66
Elson, R. M., 230
Employee Retirement Income Security Act of 1974, 164
Employment law, 161, 164
   federal, 165
   public rights and, 165
   versus labor law, 166
Endangered Species Act, 241
Entitlements, 97

Index

Environmental Bill of Rights, 245
Environmentalism, free-market, 242
Environmental management:
  debate on restructuring, 239-246
  devolution of, 239, 240-243
  elimination of, 239
  marketization of, 243
  privatization of, 239, 241-243
  restructuring since 1994, 246-250
  Sagebrush Rebellion, 242, 256
Environmental policy, U.S.:
  Congressional efforts to restructure, 247-250
  present, 234-235
  rationale for present, 235-239
Environmental Protection Agency:
  proposed spending cuts, 249
  proposed program reductions, 249
Environmental Working Group, 150, 152
Equal Employment Opportunities Commission, 172
Equal opportunity policy (U.S.), 24
Equal outcome policy (U.K.), 24
Equal Pay Act of 1963, 164
Etzioni, A., 63, 64, 65, 67, 69, 70, 71, 73

Fainstein, S., xxiv
Fairfield, R. P., 263
Fair Labor Standards Act, 164
Falk, W. A., 44
Family and Medical Leave Act (1993), 164
Family Support Act of 1988 (FSA), 193, 194
  Job Opportunities and Basic Skills (JOBS) Training Program, 194, 195
Farm Act of 1996, xix, xxiii, xxv, xxviii, xxxii
Farm Credit System (FCS), 147
Federal Agricultural Improvement and Reform (FAIR) Act, xxviii, 141, 142, 197
  agricultural credit program changes, 146-147
  Agricultural Market Transition Program, 143
  as agro-industry aid, 156-157
  as massive and temporary buyout, 148-149
  as restructuring farm policy, 143-147
  cost of, 147
  credit provisions of, 153
  dairy program changes, 144

Environmental Quality Incentive Program (EQIP), 146
expansion of environmental programs, 146
expiration date of, 149
farmers' reasons for supporting, 157-158
Farmland Protection Program, 146
Fund for Rural America, 147, 158
large farm bias in, 149-150
modification of CCC, 144
nutrition assistance program, 147Federal Agricultural Improvement and Reform (FAIR) Act (*continued*)
peanut program changes, 144-145, 151
production flexibility contracts, 143, 149, 151
research program, 147
retention of environmental programs, 146
rural development provisions, 147
small farms and, 152-155
sugar program changes, 145, 151
uneven geography of, 150-152
unfairness of, 142, 147-155
Federalist Papers, 263
Federal Mine Safety and Health Act of 1977, 164
Federal Water Pollution Control Act, 241
Fish, S., 68
Flexible dispersion of labor, 182, 184, 185
Flexible use of scale and scope, 92
Flextime, 176
Focus on the Family, 65
Food and Agriculture Policy Research Institute, 153
Forbes, Steve, 223, 257
Ford, Gerald, 222
Foucault, M., 126
Fraser, S., 265
"Freedom to Farm" Act, 141, 145, 146, 154
Freeman, R. B., 27, 29, 32
Free trade, 132, 142, 207
Friedmann, H., 154, 156, 159
Friedrichs, J., 33, 36
Fry, E. H., 25
Fukuyama, F., 65

Galbraith, J. K., 46
Galston, W., 64, 69
Gans, H., 64, 72, 182

Garber, J., 61
Garfinkel, I, 194, 198
Garreau, J., 261
Gart, A., 12, 13
Gay rights movement, 126, 128
Gelfand, M., 265
General Accounting Office (GAO), 84
General Agreement on Tariffs and Trade (GATT), 30, 142, 207, 215, 223
Geographers' Network on Politics in America, xi
Geographical mobility, 11-12, 13
Geographic redistribution of power, local capacity and, 88-93
Geographies of exclusion, 42
Geographies of power, 42
Geographies of privilege, 42
Geographies of vulnerability, 42
Gerstle, G., 265
Giddens, A., 126
Gilbert, C. C., 264
Gillespie, E., 4, 13, 97
Gingrich, Newt, 127, 247
Gintis, H., 8
Glazer, N., 194
Glendon, M., 63
Glick Schiller, N., 211, 218
Globalization, 260, 269
  economically marginal people and, 41
  economically marginal places and, 41
  economic effects of on U.S. population, 41
  economic life prospects and, 42
  economic polarization and, 42, 253
  growing income inequality and, 27-32
  regional importance and, 261
  See also Economic globalization
Glocalization, 90, 258, 260
Goldman, A. L., 163
Goldsmith, W., 43
Goldwater, Barry, 222
Goodman, D., 154, 157
Gorski, H., 98, 100, 105, 111
Gottmann, J., 212
Government downsizing, 172
Government funding, federal:
  and relationship to state/local governments, 86-87
  block grants, 86-87
  categorical grants, 86
Gramsci, A., 214, 254, 262
Gramsci's hegemony, 262-263, 265, 269
  historical blocs, 262

Gransbery, J., 257
Grant, D. S. II., 34
Gray, A. W., 159
Green, M., 93
Greenberg, M., 89
Greenberg, S., xxiv, xxviii, 67, 70
Gueron, J., 194, 198

Habermas, J., 6, 7, 252
Hall, S., 122
Handler, J., 181, 182, 183, 186, 190, 191, 194, 202
Handlin, O., 212
Harrison, B., 12
Harvey, D., 11, 51, 89, 90
Harvey, T., 34
Hathorn, S., 44
Haveman, R., 186, 194
Health care reform, xxix
Heffernan, W. D., 156
Hendrickson, M., 157
Heritage Foundation, 61, 65, 245
Herod, A., 92, 93
Hertzberg, H., 55
Hicks, D. A., 32, 34
Higham, J., 206
Hill, R., 45, 52
Hirsch, J., 6
Hobbes, Thomas, 200
Hobsbawm, E. J., 210
Hodge, Scott, 245
Hodgkinson, V., 98, 100, 105, 111
Hollinger, D. A., 210, 217
Honan, Raena, 244
Hoover, Herbert, 162
Hosansky, D., xiv
Hossfeld, K., 54
Hudson, R., 90
Hughes, J., 12, 13
Hutchison, R., 34

Identity politics, 57
Illegal aliens, 213
Immigration, illegal, 206
  increasing state efforts to stop, 208
Immigration and Naturalization Service (INS), 206, 214
  Clinton's restructuring of, 208
  funding growth of, 215
  Operation Gatekeeper, 209, 213
  Operation Hold the Line, 209, 213

Immigration Reform and Control Act
    (1986), 164
Income inequality, U.S.:
    evidence of meritocratic, 24
    geography of, 32-36
    government and, 21
    politics of, 36-38
    since 1970, 23
    See also Income inequality, growing
        U.S.
Income inequality, growing U.S.:
    deindustrialization and, 28
    deunionization and, 27, 28-29
    economic globalization and, 29-32
    evidence for, 22-27
    globalization and, 27-32
    immigration and, 31-32
    technological change and, 27-28, 31
Individualism:
    as loss of agency, 63
    in U.S. South, 266
    radical, 63
    reaction against, 62-65
Ingersoll, B., 148, 149, 151
International Union for the Conservation
    of Nature, reduction in U.S. con-
    tributions to, 250

Jessop, B., xxii, 90, 184
Job Creation and Wage Enhancement
    Act, 246
Johnson, C., 49, 50
Johnson, N. C., 210
Johnston, R. J., 255
Jones, H., 149, 158
Jones, J., 42
Jones, J. P., 44, 46, 210
Jones, J. P. III, 198
Jost, K., 240, 242

Kaestle, C., 224
Kahne, J., 229, 230
Kaplan, D. H., 34
Karger, H., 91, 92
Karpatkin, R., 50
Katz, L. F., 27, 29, 32
Kearney, M., 210, 211, 213, 218
Kemmis, D., 61, 63, 64, 65, 66, 68, 69, 73
Kennedy, P., 227
Kettering Foundation, 66, 70
Kettl, D., 84, 85, 94

Key, V. O., Jr., 266
Kilborn, P., 55, 56
Kletzer, L., 45
Klosen, S. L., 159
Kodras, J., 44, 46, 48, 54, 58, 86
Kosters, M. H., 26, 28
Kresl, P. K., 34
Krugman, P., 24, 26, 27, 28, 29
Krugman, P. R., 28
Kymlicka, W., 71

Labor:
    large capital and, 15
    service sector employment and, 16
    small capital and, 15
Labor law, 161, 164
    versus employment law, 166
Labor market flexibility, 167
Labor regulation, localization of, 162
Labor relations/regulations:
    history of modern, 162-167
    implications of restructuring, 177
    Republican agenda for restructuring,
        173-176
Lake, R., 18
Larsen, J., 54
Lauber, W. F., 229
Law, R., 44
Lawrence, R., 29
Lawrence, R. Z., 28
Leamer, E. E., 26, 28, 30, 31, 33, 34, 37
Lefebvre, H., 126, 127
Left, 118
    rights-talk by, 119, 120
Legitimation role of state, xxiii, xxiv, 6,
    252, 256, 257, 158, 269
Le Grand, J., 83
Lehmann, S., 240, 241, 242, 243, 251
Leitner, H., 92
Leuchtenberg, W. F., 222
Levine, M. V., 26, 34, 36
Lie, J., 211
Lipsky, M., 110
Lugar, Richard, 141, 148
Lukes, S., 8
Luria, D., 52

MacKinnon, C., 68
Madison, James, 64, 68
Madsen, R., 66
Magdoff, H., 35

Mair, A., 14, 89, 90, 92
Malbin, M., 86, 87
Manifest Destiny, 211, 216
Manski, C., 194, 198
Marble, Gary, 244
Margolis, D., 12
Marketization, 167
Martinez, O. J., 211
Marx, Karl, 122
Massey, D., 42, 89, 126
McDonaugh, E., 68
McMichael, P., 156
Meckstroth, D., 12
Medicare, 56
Meritocracy, myth of, 185
Mertz, P. E., 264, 265
Meszaros, I., 123
Miller, B., 91, 92, 93
Milliken v. Bradley, 228
Mine Safety and Health Act, proposed repeal of, 175
Mine Safety and Health Administration, proposed dismantling of, 175
Mining Act (1872), 248
Miranda, L., 49, 50
Mishel, L., 10, 16, 26
Mitchell, D., 125, 127, 137
Moe, T., 229
Mollenkopf, J., 33
Montreal Protocal Fund, reduction in U.S. contributions to, 250
Morici, P., 31, 34
Morone, J., 71, 73
Mouffe, C., 89
Mountz, A., 213, 218
Moynihan, Daniel Patrick, 49, 111
Musselwhite, J., 83
Myhre, D., 156

NAFTA, 34, 207, 209, 215, 223, 256
  Borderlands and, 261
  opposition to, 37, 132
Nathan, R., 87
National Biological Survey, 245
National Endowment for the Arts and Humanities, 111
National Environmental Policy Act, 249
Nationalism, 206, 216
  patriotism and, 214
National Labor Relations Act (NLRA), 163, 164, 168, 169

National Labor Relations Board (NLRB), 163, 165, 170, 171, 172
National Oceanic and Atmospheric Administration:
  research funding cuts, 249
National Park Service, 245
National Recovery Act, 264
National state control:
  of destructive competition, 84
  of fiscal redistribution, 84
  of interstate issues, 84
  of uniform standards, 84
Natter, W., 210
Natural Resources Conservation Service, 146
Natural Resources Defense Council, 4, 13, 249
Negotiated Rulemaking Act (1990), 172
Negotiated rule-making (reg-neg), 173
  as devolution, 173
  definition of, 172
New Deal, xxxiii, 9, 194, 222, 255, 257, 259, 264, 265, 267
New Deal labor laws, 163, 166
  state restructuring and, 163-164
New Federalism, 167
New Right, 118, 128, 129
Nightingale, D., 186, 194
Nijman, J., 32
Nixon, Richard M., 128
Nixon administration, 86, 87
  Family Assistance Program, 194
Noga, S. M., 98, 100, 105, 111
Nonprofit sector, 98-99
  activities, 99
  Bush administration challenges to, 98
  employment in, 100
  geographic patterns/trends in, 102-103
  growth of, 103
  locations of, 102
  national income from, 100
  need for, 104-105
  Reagan administration challenges to, 98
  revenue sources, 99, 100
  service activities, 99
  See also specific charities and non-profits; Charities and charitable agencies
Nonprofit sector capacity, 103-109
  recent social/demographic shifts and, 107-109
Noonan, P., 65

Index

Nord, S., 16
Norris-La Guardia Act, 162-163
Nygren, T. I., 104

Occupational Safety and Health Act of 1970, 164, 174, 175, 176
Offe, C., 7, 90
O'Loughlin, J., 23, 33, 36
Omnibus Budget Reconciliation Act (1981), 194
Omnibus Property Rights Act, 248
104th Congress, 231, 259
  attempts to weaken Environmental Protection Agency, 222
  domestic policy issues and, 221-222
  political agenda of, 252
  U.S. South and, 266
  See also Republican dilemma
O'Neill, M., 99
O'Neill, P., 12
Ong, A., 207, 214, 215, 216
Organizational mobility, 11, 12, 13
Osborne, D., 84
Ostrander, S., 101
Outlaw, J. L., 159

Padavic, I., 48
Pagano, M., 87
Page, B., 154, 157
Panagariya, A., 37
Paris, D. B., 224
Pateman, C., 200
Patriotism, 214, 217
Pauly, E., 194, 198
Pearce, D., 46
Pearlstein, S., 23, 24
Peck, J., 81, 89, 90, 91, 92, 162, 177, 182, 183, 184, 185, 186, 191, 200
Pedersen, W. F., Jr., 236
Personal Responsibility and Work Opportunity Reconciliation Act (PRWOR), 181, 192, 198
Peterson, P., 198
Peterson, P. E., 84
Phillips, K., 38, 50
Pine Ridge reservation, economic restructuring and, 54-56
Pines, B., 83
Piven, F. F., 183, 184, 187, 192, 201
Place context, xxxi-xxxiii
Pope, Carl, 244

Popper, F., 89
Presidential campaigns, 36
President's Council on Environmental Quality, budget reduction in, 249
Press, E., 129, 130, 137
Pressman, J., 86
Prestowitz, C., 29
Private Property Owners' Bill of Rights, 240, 245
Privatization, xxviii, 80, 83, 87, 89, 167, 196, 253, 256, 269
  as spatial process, 82
  definition of, 81
  examples of, 81
  for-profit sector, 81
  nonprofit sector, 81-82
  of labor-management relations, 162, 173
  redefining government's scope, 82
  welfare and, 184
Productivity, 26
Promise Keepers, 268
Public Interest Research Group (PIRG), 245
Public-private partnership, xxiv
Putnam, R., 63, 201

Quality of Worklife Circles, 176

Ramey, V. A., 28
Rankin, B. H., 44
Rauber, P., 238, 244, 245
Reagan, Ronald, 128, 222, 230
Reagan administration, 86, 87, 194
  budget slashes, 202
  education reform and, 229
  social policy cutbacks of, 55-56
Reagan democrats, 16
Reardon, L., 12
Recession of 1990-1991, 10-11
Red Cross, 112, 116
Re-engineering, 26
Regulatory Flexibility Act, 246
Reich, Robert, 167, 172
Republican dilemma:
  and effects on education policy, 226, 231
  education and, 224-225
Republican National Committee, 67
Reynolds, D. R, 224, 225, 227
Richardson, J. W., 159
Ridgeway, C., 242

Rights:
  as process of producing space, 127
  institutionalization of, 126
  versus moral obligations, 64
Rights-talk:
  as necessary, 123
  institutional power and, 125
  Rorty's criticism of, 119-120
  social movements and, 123
  Tushnet's criticism of, 121
Roberts, Pat, 141, 142, 148, 149, 150
Robinson, K., 85
Robinson, R., 83
Rodriguez, R., 212
Rogers, E., 54
Romer v. Evans, 119, 125, 128, 138
Roosevelt, Franklin D., 128, 163, 222, 264
Rorty, R., 64, 71, 119, 120, 122, 136
Rosecrance, R., 265
Rosenberg, T., 44
Rouse, R., 206, 207, 210, 213, 214, 215
Rousseau, Jean Jacques, 200
Rupert, M. E., 256, 260
Russell, J., 52

Sack, R., 210
Safety and Health Improvement and Regulatory Reform Act, 174
Salamon, L., 83
Salamon, L. M., 111
Salvation Army, 99, 112, 116
Samuelson, Robert, 23
Sandel, M., xiii, 60, 61, 63, 70, 74
Santa Clara County v. Southern Pacific Railroad Co., 131, 138
Saxenian, A., 53
Schattschneider, E. E., 253, 255
Schellhas, B., 4, 13, 97
Schlesinger, A. M., 38
Schmidt, John R., 240
Scholssstein, S., 227
Schools, public:
  curricular bias in, 230
  multicultural curricula in, 230-231
  See also Common schools
Schwab, R., 83
Schwartz, A., 34
Seabrook, J., 42
Sectoral mobility, 11, 12, 13
Seneca, J., 12, 13
Shadow state, nonprofit sector as, 82
Sheets, R., 16
Shelley, F. M., 224, 225, 227, 229

Shepperson, W. S., 212
Sherman, A., 49, 50
Shklar, J., 71
Sibley, D., 42
Sierra Club, 244, 245, 246
  Environmental Rights Activist Network, 246
Silicon Valley, economic restructuring and, 53-54
Silvern, S., 92
Skocpol, T., 192, 193, 194
Slaughter, M., 29
Small businesses, local dependence of, 14-15
Smith, C., 83
Smith, D. L., 265
Smith, E. G., 159
Smith, M. P., 214
Smith, N., 92, 126, 129
Smith, S. R., 110
Smith, T. T., 235, 239
Social action, 127
Social composition of local population, 51
Social contract theorists, 200
Social Darwinism, failure of, 97
Social Darwinists, 244
Social movements, 91
  power of local, 91
  See also specific social movements
Social restructuring in American population, 43-44
  gender and family composition and, 46-48
  generations and, 49-50
  political implications of, 56-57
  race/ethnicity and, 44-46
Social Security, 56
Social Service Block Grant, 111
Soguk, N., 211
Soja, E., 126
Solo, R., 6
Sorj, B., 154, 157
South, U.S.:
  collective memory of, 265-266, 267
  colonial economy of, 264
  ideology of, 267
  individualism and, 266
  morality theme and, 267
  positioned as source of values, 269
  pro-accumulation rhetoric of, 268
  responsibility theme and, 267
  states' rights and, 266
  unique regional history of, 263-264
Southern League, 267

Index

Southern State Industrial Council (SSIC), 264
Spatial organization of the United States, xxx
Spatial scale, xxx
Spiegel, J., 12, 13
Stacey, J., 54
Staeheli, L., 66, 67, 68, 70, 71, 267
State, xxxiii, xxxiv
  as key protector of the weak, 123
  capital and, xxiv
  definition of, xxii, xxiii
  functions of, 6
  local, 90
  national, 90
  See also Accumulation; Legitimation
State actors, 8
State autonomy, relative, 5, 6-10
  theory of state derivation and, 6
State change:
  implications of, xv
  processes of, xv, xxiii, xxxiv
  sources of, xv
  See also State restructuring
State restructuring, 17-18, 252
  discourse of, xxvii-xxix
  extent of, 9
  forms of, xxviii, 17
  influences on, 5
  new geographies created by, xxx-xxiii
  politics of, 255-258
  strategies, 79, 80-88
  See also Devolution; Dismantling; Privatization
Stoesz, D., 91, 92
Storper, M., 90
Strange, M., 149, 154, 159
Strohbehn, E. L., 234, 236
Sullivan, M., 66
Swidler, A., 66
Swyngedouw, E. A., 90, 258, 260
Szanton Blanc, C., 211, 218

Takaki, R., 212
Taney, Roger, 131
Task Force on Persistent Rural Poverty, 56
Tax Reform Act of 1986, 115
Taylor, P. J., 261
Teamwork, 176
Teamwork for Employees and Managers Act of 1995, 168
  provisions of, 169-170
Thelen, K. A., 168

Thomas, J., 45
Thomas, R., 45
Thomas, W. I., 212
Thomma, S., 68
Tickell, A., 90
Tienda, M., 44, 45
Tipton, S., 66
Trans-local alliances, 93
Transnationalism, 206, 210-211, 216
  diverse societies and, 215-216
Transnational/national dialectic, 206, 210
Tucker, A. M., 229
Turner, S. E., 104
Tushnet, M., 120, 121, 122, 124, 125, 126, 130
Tyack, D., 225, 226

U.S. Border Patrol, 206
  restructuring, 208
U.S. Bureau of the Census, 31, 45, 47, 49, 209, 216
U.S. Department of Energy:
  budget cuts, 249
U.S. Department of Interior:
  ban on information collection of threatened or endangered species, 249
U.S. Fish and Wildlife Service, 245
U.S. Forest Service, 245
U.S. Geological Survey, 245
U.S. House of Representatives, 84
Uehlein, 130
Unger, R., 130
Unions, 134
  social movements and, 129
United Nations Environmental Programme, reduction in U.S. contributions to, 250
United States Department of Agriculture (USDA), 145, 147, 154
  Farm Service Agency (FSA), 146, 153, 154
United Way, 92, 112-113
Urrea, L. A., 212
Uruguay Round Agreement (URA), 142
USDA, 148, 157, 159

van Wezel Stone, K., 163
Venables, A. J., 28
Voluntarists, 83
  critics of, 84
Voluntary organizations, local, 91

Wagner Act, 165. *See also* National Labor Relations Act (NLRA)
Waite, Morrison, 131
Waldron, J., 125
Walker, R., 34, 90, 154
Wallerstein, I., 23, 254, 259, 260
Warf, B., 89
Watts, M., 157
Weill, J., 49, 50
Weitzman, M. S., 98, 100, 105, 111
Welfare, devolution of, 181, 184, 196-199, 203
  recent trends in, 192-196
Welfare, discourse of devolution of, 185-191
  gender and racial significance of, 187, 190
  poverty and, 190
  primacy of focus on welfare programs and, 191
  regional and local contexts and, 190
  workhouse regions and, 191
Welfare, dismantling, 201-203
Welfare, privatization of, 199-201, 203
"Welfare fatigue," 36
Welfare programs:
  as target for restructuring, 182-185
  emergence of modern, 97
  See also Welfare reform; Workfare
Welfare reform, 181
  state-level, 186
  workfare legislation and, 181
Welfare Reform Act of 1996, xx, xxvi, xxxii, 25, 199
West, B., 89
Wetlands Reserve Program:
  retention of in FAIR Act, 146
White, T. H., 222
WIC, 56

Wiecek, W., 131
Wildavsky, A., 86
Wilkinson, J., 154, 157
Williams, P., 136
Wilson, C., 45, 52
Wilson, Pete:
  anti-immigration platform of, 208
Wilson, W. J., 58
Wilson, Woodrow, 86
Wilterdink, N., 35, 36
Wolch, J., 44, 82, 91
Wolfe, A., 6
Wolpert, J., 84, 88, 98, 102, 106, 116
Worker Adjustment and Retraining Notification Act (1988), 164
Workfare, 72, 184
  evolution of, 194
Workfarist state, 184, 185
Workplace law. *See* Employment law; Labor law
Works Progress Administration, 194
World Economic Forum, 26
World Meteorological Organization, reduction in U.S. contributions to, 250
World Trade Organization (WTO), 30, 37, 133
World War II, 86
Wright, R., 210, 213, 218
Wyman, M., 212, 213

Young, Don, 244

Znaniecki, F., 212
Zopf, P., 46
Zukin, S., 42

About the Contributors

MEGHAN COPE is Assistant Professor in the Department of Geography at the State University of New York at Buffalo. Her interests lie in the areas of the geographic perspectives on gender and race, social policy, labor and employment, and the challenges faced by deindustrialized cities of the northeastern United States.

COLIN FLINT is Assistant Professor in the Sam Nunn School of International Affairs, Georgia Institute of Technology. He is a political geographer whose interests include electoral geography, international political economy, and world-systems theory. His published research includes articles on the electoral geography of the Nazi party, English nationalism, and world-systems theory in the *Annals of the Association of American Geographers, Political Geography,* and *Sociological Inquiry.*

ANDREW HEROD is Associate Professor of Geography at the University of Georgia, at Athens. His research interests focus on the geography of labor and trade unionism both in the United States and abroad. He is currently conducting research on how the transition in Eastern Europe is affecting trade unions and the geopolitics of labor in this region. He is editor of *Organizing the Landscape: Labor Unionism in Geographic Perspective* (forthcoming) and co-editor with Gearóid Ó Tuathail and Susan Roberts of *An Unruly World? Globalization, Governance and Geography* (1997).

JANET E. KODRAS is Professor of Geography at Florida State University. She is co-chair of the Geographers' Network on American Politics, a nationwide consortium of scholars seeking to inform the public of the great geographical complexities of governance in the United States. Her research focuses on the spatial dimensions and dynamics of social policy with particular attention to poverty and hunger. She is the author of numerous studies on the place- and time-contexts of social problems and the welfare state and is co-editor with John Paul Jones of *Geographical Dimensions of U.S. Social Policy.*

ROBERT W. LAKE is Professor at the Center for Urban Policy Research and a member of the Graduate Faculty in the Department of Geography and the Department of Urban Planning and Policy Development at Rutgers University. He is a member of the Steering Committee of the Geographers' Network on American Politics. He is the author of numerous books and articles on state theory, local autonomy, environmental policy and environmental justice, and political conflicts over siting unwanted facilities.

DON MITCHELL is Assistant Professor in the Department of Geography, at the University of Colorado. He is the author of *The Lie of the Land: Migrant Workers and the California Landscape.* His research interests include the geography of political struggle in public space, the historical geography of labor and landscape, and the political economy of "culture."

JOHN O'LOUGHLIN is Professor of Geography and Faculty Research Associate in the Institute of Behavioral Science at the University of Colorado. He is a member of the Program on Political and Economic Change's graduate research training program on "Globalization and Democratization," funded by the National Science Foundation. His current research interests are in the diffusion of democracy, the economic and political transition in the countries of Central Europe and the former Soviet Union, regional protest parties in Europe, and the growing social polarization of European cities in an era of globalization. He is North American editor of *Political Geography.*

BRIAN PAGE is Assistant Professor of Geography at the University of Colorado at Denver. His major research interests are economic geography, historical geography, and political economy. He has written articles on the meat packing industry, historical regional development in the Midwest, and the restructuring of U.S. agriculture.

FRED M. SHELLEY is Associate Professor of Geography at Southwest Texas State University. His research interests include political geography, electoral geography and the United States. He is author of several books including *Political Geography of the United States.* He is Review Editor of *Social Science Quarterly* and *Political Geography.*

About the Contributors

LYNN A. STAEHELI is Associate Professor of Geography and Faculty Research Associate at the Institute of Behavioral Science at the University of Colorado. She is a member of the Program on Political and Economic Change's graduate research training program on "Globalization and Democratization," funded by the National Science Foundation. She is also a member of the Steering Committee of the Geographers' Network on American Politics. Her current research examines changes in local democratic institutions and practices, communities, citizenship, gender, immigration, and social service delivery.

MARVIN WATERSTONE currently holds joint appointments as Associate Professor of Geography and Regional Development and Associate Professor of Hydrology and Water Resources at the University of Arizona. In addition, he is the Director of the Graduate Interdisciplinary Program in Comparative Cultural and Literary Studies. His current interests focus primarily on the intersections of discourse and power, and the uses of social theory. In addition, he maintains an active research and teaching program in issues of environmental and resource management, and in the human dimensions of global change. In this connection, Dr. Waterstone is currently Associate Editor for the *Annals of the Association of American Geographers* for Nature/Society Relations.

JULIAN WOLPERT is the Henry G. Bryant Professor of Geography, Public Affairs, and Urban Planning at Princeton University's Woodrow Wilson School. He is a member of the Steering Committee of the Geographers' Network on American Politics. His current teaching and research is focused on U.S. domestic policy relating to service provision in metropolitan areas. He chairs the Independent Sector's Research Committee and previously chaired the National Center for Charitable Statistics. His most recent nonprofit study, *What Charity Can and Cannot Do,* was published by the Twentieth Century Fund in 1996.

RICHARD WRIGHT received his undergraduate degree from the University of Nottingham in England and his graduate degrees in Geography from Indiana University. He has been in the Geography Department at Dartmouth College since 1985 and is currently Professor and Departmental Chair. His main research interest, supported by grants from the National Science Foundation and Social Science Research Council, involves the analysis of the linkages between international and internal systems of labor exchange, and more generally, the question of how immigrants to the United States fit into the metropolitan and nonmetropolitan economies of the country. He is also interested in the transnational migration of labor and capital and how to understand the international movement of people, goods, information, and money. His most recent publications have appeared in *Economic Geography* and *Urban Geography*.